STELLAR PATHS

ASTROPHYSICS AND SPACE SCIENCE LIBRARY

A SERIES OF BOOKS ON THE RECENT DEVELOPMENTS
OF SPACE SCIENCE AND OF GENERAL GEOPHYSICS AND ASTROPHYSICS
PUBLISHED IN CONNECTION WITH THE JOURNAL
SPACE SCIENCE REVIEWS

VOLUME 85

PETER VAN DE KAMP

STELLAR PATHS

PHOTOGRAPHIC ASTROMETRY WITH LONG-FOCUS INSTRUMENTS

with an Introduction by Jean-Claude Pecker

D. REIDEL PUBLISHING COMPANY

DORDRECHT : HOLLAND / BOSTON : U.S.A.

LONDON : ENGLAND

Library of Congress Cataloging in Publication Data

CIP

van de Kamp, Peter, 1901–
Stellar paths.

(Astrophysics and space science library; v. 85)
Bibliography: p.
Includes index.
1. Astrometry. I. Title. II. Series.
QB807.V33 523.8′1 81–8505
ISBN-13: 978-94-009-8452-3 e-ISBN-13: 978-94-009-8450-9
DOI: 10.1007/978-94-009-8450-9

Published by D. Reidel Publishing Company,
P.O. Box 17, 3300 AA Dordrecht, Holland.

Sold and distributed in the U.S.A. and Canada
by Kluwer Boston Inc.,
190 Old Derby Street, Hingham, MA 02043, U.S.A.

In all other countries, sold and distributed
by Kluwer Academic Publishers Group,
P.O. Box 322, 3300 AH Dordrecht, Holland.

D. Reidel Publishing Company is a member of the Kluwer Group.

to Maja

TABLE OF CONTENTS

The Sproul refractor in Swarthmore, Pennsylvania with Dr John L. Hershey. Aperture 61 cm, focal length 10.93 m. Scale in focal plane 1 mm = 18.87 or 1″ = 53 microns.

CONTENTS

PREFACE

This is the latest effort in a sequence of presentations begun in 1949 with a series of lectures on long-focus photographic astrometry given by the author as Fulbright professor in Paris at the invitation by the late H. Mineur, at that time Director of the Institut d'Astrophysique. These earlier lectures were published as a series of review articles in *Popular Astronomy* (1951) and appeared both as *Contributions de l'Institut d'Astrophysique*, Série A, No. 81 and as reprint No. 75 of Sproul Observatory. A more elaborate presentation was given in 1963 in *Stars and Stellar Systems*, which was followed by *Principles of Astrometry* (1967, W. H. Freeman & Co.).

During the second half of 1974, again under Fulbright auspices, at the invitation of Pik Sin The, I lectured at the Astronomical Institute in Amsterdam, followed by a short course in May–June 1978 at the invitation of E. P. J. van den Heuvel. I gave a more extensive course at the Institut d'Astrophysique at the invitation of J. C. Pecker of the Collège de France and of J. Audouze, Director of the I.A.P. Both in Amsterdam and in Paris I had presented occasional astrometric topics at various times. The opportunity to lecture in France and in Holland has facilitated, influenced and improved the organization and contents of the presentations on the subject of long-focus photographic astrometry.

Since the publication of *Principles of Astrometry* a number of review articles and research articles by the author have appeared in the *Annual Review of Astronomy and Astrophysics, Vistas in Astronomy, The Astronomical Journal, Sky and Telescope, et al.* These together with the results of investigations by others are the principal source and *raison d'être* for the present update and sequel to Part II of *Principles of Astrometry*. For the sake of completeness and continuity there is unavoidable overlap with the earlier book. In these cases repetition has frequently been abbreviated and improved, and reference is made to earlier presentations. Unless otherwise stated, probable errors are used.

Inevitably the results of the present effort are not complete, they are selective and are to a great extent a narrative of those topics with which the author has been involved. Following a brief survey of historical highlights (Chapter 1), this treatise deals with the principles, technique and accuracy of the study of individual stellar paths, differentially observed on a background of reference stars within small angular distance (Chapters 2, 3, 4). The results for individual stars yield information about parallax, proper motion and quadratic time effect. The reduction from relative to absolute values of stellar accelerations is discussed in some detail (Chapters 5, 6, 7, 8, 9). Next follows a review of double-star kinematics, orbital motion, and mass determinations (Chapters 10, 11, 12).

Particular attention is paid to the study of perturbations in stellar paths due to unseen astrometric companions, this latest and promising branch of long-focus photographic astrometry; selected illustrations and a brief review are given (Chapters 13, 14, 15, 16).

The last two chapters deal with long-period eclipsing binaries (Chapter 17), and with attainable accuracy, sub-stellar mass and planetary detectability (Chapter 18).

I am indebted to the Institut d'Astrophysique, the Collège de France, the University of Amsterdam as well as the National Science Foundation, the Netherlands Foundation for Zuiver Wetenschappelijk Onderzoek (ZWO), the Leids Kerkhoven–Bosscha Fonds, and the Fulbright–Hays Commission, for their faith in me and their continuing interest in the fundamental field of Astrometry. My thanks also to the staff members of Sproul Observatory, above all to Sarah Lee Lippincott and John L. Hershey for assistance in preparing the book, to Robert S. Harrington, Wilhelm Gliese and many others for valuable suggestions and material. Thanks are due to staff members of the Astronomical Institute in Amsterdam, especially to Ed Faverey for his painstaking preparation of illustrations; and to Kjongae Chang of the Hamburg Observatory for her assistance in the final stages of the manuscript.

The author also acknowledges the kind permission to reproduce 15 illustrations and portions of the text published in *Principles of Astrometry*, W. H. Freeman & Co., San Francisco and London.

1980 December P.v.d.K.
Swarthmore, Pennsylvania
Amsterdam, The Netherlands

INTRODUCTION

When, a few decades ago, the author of these lines was preparing to enter upon a career in astrophysics, his director at the Paris Observatory and the CNRS, André Danjon, advised him (in the imperative mode, to be sure) to spend at least a little time in the hard school of astrometry, as practiced in the Meridian Service of the Observatory of Paris.

Of course I followed such persuasive counsel. No doubt for this reason did I then become, irredeemably, an astrophysicist! That was in 1946. A few years later, in 1949 (if I remember correctly), another type of astrometry occupied me for several weeks: I attended the first course that Peter van de Kamp gave at the Institut d'Astrophysique, on long-focal-length photographic astrometry.

It was the time when the new astrophysics strode the scene as conquering hero, when reflectors were replacing refractors, when the positions or motions of the stars seemed of petty import compared to the problems of their physical nature.

However, for some of us, van de Kamp's course, evoking a science old perhaps in its basic concepts, classical no doubt in its techniques, nonetheless appeared to foreshadow the instrumental development necessary for the astronomy of the 21st century, and the philosophical concerns of the scientists (astronomers and others) of tomorrow. How wrong to separate the domains of astrometry and of astrophysics by the artificial barriers of tradition or ambition! Is it even thinkable to speak of knowing one without the other, or rather to attempt to practice one without a good knowledge of the other? Furthermore, between rockets and infrared detectors, between quasars and flare stars, what has become of unity of this 'astrophysics'? 'Astrophysics', whose heralds regarded 'astrometry' as one might take a look around a dusty museum, amusedly musing to be sure, but with a trace of scorn for what was obviously a vanishing species.

It would seem that astronomy has become possessed with the ecological spirit, and that the vanishing species of astrometry, so well protected by enthusiasts like Peter van de Kamp, has made such a success of survival as to impose itself as one of the most lively and fertile disciplines of contemporary astronomy. The intuitions suggested by van de Kamp's 1949 lectures have become certainties. And this, I find, is a perfectly valid excuse for me to have the pleasure of prefacing a treatise on long-focal-length astrometry, having already prefaced Paul Couteau's work on double stars. I am moreover delighted and honored to preface the book of the same van de Kamp whose lectures I humbly listened to in 1949, and who now asks this preface of me as a friend, although he certainly has no illusions regarding the state of my incompetence in the domain over which he reigns.

Long-focal-length astrometry, with its proven techniques, repeated over years, repeated over decades, yields the apparent motions of the observable stars with remarkable precision (a hundred-thousandth of a second of arc in the linear terms of the annual proper motion, a millionth of a second of arc in the quadratic terms). Double and multiple star systems are of course used for mass determinations (and where would theoreticians of stellar evolution be without masses?). From parallaxes, distances are determined, and then luminosities... And again, without them, what could theoretical astrophysicists prove?

These fundamental results remain limited, however. In his book, van de Kamp shows how, by prolonging the period of observation, the data can be improved, so that proper accelerations as well as proper motions can be determined, so that a volume a thousand times greater than the presently known sphere of a few hundred parsecs surrounding the Sun can be explored. Very well. But it is clear that much remains forbidden to us: with the instruments used by van de Kamp, for all except the closest stars, limitations arise because the objects to be studied are too faint, their proper motions too small; moreover there are difficulties in defining an absolute system of reference. At the same time, we know (certainly better than 50 years ago) what is needed to progress. Complete understanding of galactic dynamics is indispensible, together with a detailed exact study of the complex rotation of our Galaxy, if only to arrive at a correct evaluation of its mass, a basic astronomical quantity. Precise determinations of distances (hence of luminosities) of stars of all types are absolutely necessary, including nearby faint stars (pulsars, white dwarfs, ...), including bright but very distant stars (supergiants, novae, ...) as well. Peter van de Kamp argues vigorously for the maintenance of long-focal-length refractors, for the uninterrupted pursuit of their programs, and he is right, without a doubt. But, without trying to predict what the astrometry of the end of the century will be, it is clear in the light of his detailed discussion, that extraterrestrial instruments will be needed – astrometric satellites, loudly called for today by both astrometrists and astrophysicists, in harmonious concert.

Prudently, and modestly, Peter van de Kamp introduces us to a completely different domain. Thanks to his efforts, invisible bodies have been detected, through their perturbations of the proper motions of visible stars. What kind of bodies? Are they faint but still stellar companions? Or planets in the usual sense of the term? (And is Jupiter a planet in the usual sense? Its long-wave emission suggests not . . .). Van de Kamp's discovery, a few decades ago, created a sensation in the world of science, surprised and delighted with the discovery of new planetary systems. Some disappointment has perhaps set in since then: in recent years, the information on the properties of known systems has been refined – in particular the companion-to-star mass ratio – but the list of possible solar systems has hardly grown. And yet, such growth was hardly likely, for the instruments remained the same; and one could hardly expect to detect many more nearby objects perturbed by invisible masses. The (completely independent) discovery of binaries with massive (stellar if you like) but compact (black hole?) companions, throws a new light on this problem. So do the attempts at a statistical evaluation of the number of planetary systems in the Universe, based on considerations of galactic evolution. Van de Kamp's significant examples, few because very nearby, constitute the one solid argument of this last development, giving a firm foundation to more speculative research, placing reasonable constraints on such conjecture.

Planets, here or there, and not only around our Sun? Certainly. Peter van de Kamp has shown the existence of a few very well established cases, and this is fundamental. And on such planets, Life? That is a completely different question; and the probabilities introduced into many classical discussions of it are for the most part but the barest of intuitions. Life? . . . Maybe . . . And it is not madness to search for it; OZMA, SETI, and other such projects have seen fine days, and can look ahead to many more. We shall see.

In any case one thing is certain. Whether we seek to explore methodically the nearby universe and to know its dynamical properties, or whether we seek to detect other solar systems than our own – infinite patience is necessary. It is such patience that is the dominant virtue of the astrometrists, of those who like van de Kamp devote a lifetime of work and their immense talent to improving the precision of the measurements and to extracting a few rare results – rare but sure. Such patience betrays a passion . . . With such peers, we astrophysicists cannot fail to be encouraged. Truly enough, many sensational discoveries, as stirring as they may be, do not revolutionize science; and the pressing need remains for the exemplary precision which over the centuries has been the joy and the prestige of astronomers.

February 1980

JEAN-CLAUDE PECKER
(Translated by Robert S. Kandel)

INTRODUCTION

Lorsqu'il y a quelques décennies, l'auteur de ces quelques lignes se préparait à entamer une carrière d'astrophysicien, André Danjon, qui l'avait accueilli à l'Observatoire de Paris et au CNRS, lui conseilla (et ce conseil avait, n'en doutons pas, des vertus impératives) de pratiquer, au moins un temps, l'ecole de l'astrométrie rigoureuse du Service Meridien de l'Observatoire de Paris.

Le conseil persuasif, je l'ai suivi bien sûr. C'est sans doute pour cela que je suis alors devenu astrophysicien, irrémédiablement! C'était en 1946. Peu d'années plus tard, en 1949, si je me souviens bien, une autre espèce d'astrométrie m'occupa quelques semaines: je suivis, à l'Institut d'Astrophysique, les premiers cours qu'y donna Peter van de Kamp, sur l'astrométrie photographique à long foyer.

C'était l'époque où la jeune astrophysique se posait en héroïne, en conquérante, où le télescope concurrençait la lunette, où la position des astres, voire leur mouvement, semblait sans grand intérêt au regard de leur nature.

Pourtant, le cours de van de Kamp éveilla, au contraire, chez plusieurs d'entre nous, le sentiment d'une science qui, sur des concepts peut-être anciens, avec des techniques sans doute classiques, préfigurait à la fois les développements instrumentaux necessaires de l'astronomie du XXIème siècle, et les préoccupations philosophiques des scientifiques (astronomes ou autres) de demain. Quelle erreur de séparer par les artificielles cloisons des traditions ou des ambitions les deux royaumes, astrométrie, et astrophysique! N'est-il pas devenu impensable de connaître l'un sans l'autre, ou plutôt, de pratiquer l'un sans une bonne connaissance de l'autre? Et d'ailleurs, au milieu des fusées, des récepteurs infrarouges, des quasars et des étoiles à éruptions, qu'en est-il donc de l'unité de cette astrophysique dont les hérauts jetaient naguère à 'l'astrométrie' le regard de celui qui parcourt un musée poussiéreux, amusés à muser bien sûr, mais quelque peu méprisants pour cette espèce de toute évidence en voie de disparition?

Il faut croire que l'âme écologiste a soufflé sur l'astronomie, et que l'espèce en voie de disparition, bien protégée par les enthousiastes comme Peter van de Kamp, a réussi sa survie au point de s'imposer comme l'une des disciplines les plus vivantes, les plus prometteuses de l'astronomie contemporaine. Les intuitions que nous avaient suggérées les cours de 1949 sont des certitudes aujourd'hui. Et ceci est, je pense, une excuse très valable à ce qu'après avoir préfacé un ouvrage de Paul Couteau sur les étoiles doubles, je me sente heureux, cette fois-ci encore, de préfacer un ouvrage sur l'astrométrie à long foyer; heureux et honoré que ce livre soit celui du même van de Kamp dont j'écoutais humblement les leçons en 1949, et qui me fait aujourd'hui l'amitié de me le demander, bien qu'il ne se fasse, c'est clair, aucune illusion sur mon incompétence notoire dans le domaine où il règne.

L'astrométrie à long foyer, les techniques éprouvées, répétées des années durant, des décennies durant, permettent d'atteindre, avec des précisions remarquables (un cent-millième de seconde de degré sur les termes linéaires du mouvement propre annuel, un millionnième de seconde de degré sur le terme quadratique de ce mouvement) les mouvements apparents des étoiles observables. Les étoiles doubles et multiples sont bien sûr utilisées à la détermination des masses (et que feraient donc sans cela les théoriciens de l'évolution des étoiles?). Les parallaxes fixent les distances, donc les luminosités... Et là encore, sans cela, que feraient les théoriciens de l'astrophysique?

Ces résultats fondamentaux sont cependant limités. La lecture de l'ouvrage de van de Kamp montre certes que la prolongation des observations permet d'améliorer les données, d'atteindre les accélérations propres et plus seulement les vitesses propres, d'explorer un volume qui ne sera plus seulement la sphère de quelques cent parsecs qui entoure le soleil, mais une sphère sans doute mille fois plus volumineuse. Soit. Mais il est clair que bien des choses restent interdites; les limitations imposées, s'il ne s'agit pas d'étoiles proches, par l'éclat des objets étudiés et par la petitesse de leurs mouvements propres, la difficulté qu'il y a à définir parfois les systèmes absolus de référence, tout cela impose des limitations à l'instrumentation utilisée par van de Kamp. En même temps, on sait (mieux sans doute qu'il y a cinquante ans) ce dont on a besoin. Une connaissance précise de la dynamique galactique est indispensable, une étude parfaite de la rotation complexe de notre Galaxie, et ce, ne serait-ce que pour parvenir à une évaluation sans erreur de sa masse, quantité fondamentale de l'astronomie. Une détermination précise des luminosités (donc des distances) des étoiles de tous types est indispensable, même des étoiles proches et faibles (pulsars, naines blanches...), même des étoiles brillantes mais très lointaines (supergéantes, novae). Peter van de Kamp plaide avec vigueur pour le maintien des réfracteurs à long foyer, pour la continuation ininterrompue de leurs programmes. Il a raison, à n'en pas douter. Mais, sans préjuger de ce que deviendra l'astrométrie de la fin du siècle, il est clair, à la lumière de la discussion détaillée, qu'elle devra aussi faire usage d'engins extraterrestres, de ces satellites astrométriques, que réclame aujourd'hui à grands cris l'harmonieux concert des astrométristes et des astrophysiciens.

C'est aussi dans un tout autre domaine que nous entraîne, avec prudence et modestie, Peter van de Kamp. C'est grâce à ses travaux qu'autour d'étoiles visibles, et dont le mouvement propre est perturbé, la détection d'astres invisibles est acquise. De quels astres s'agit-il? Compagnons faibles, mais encore stellaires? Ou planète au sens usuel du terme (et d'ailleurs Jupiter est-il une planète au sens usuel? Son rayonnement infra-rouge semble bien démontrer le contraire)...La découverte de van de Kamp, il y a quelques décennies, remue le monde scientifique, étonné et ravi de la découverte de nouveaux systèmes planétaires. Une sorte de déception s'est peut-être faite jour depuis lors: dans les années récentes, on a précisé les caractéristiques des systèmes connus – notamment le rapport de la masse du système compagnon à celle de l'étoile – mais on n'a que peu allongé la liste des systèmes solaires possibles. Pourtant, un tel progrès n'était guère probable car les instruments restaient les mêmes; et quelques années de plus ne pouvaient, on le savait bien, modifier sensiblement le nombre d'objets proches détectés comme perturbés par des masses invisibles. La découverte (de façon totalement indépendante) de binaires dont un compagnon est massif – stellaire si l'on veut) mais nain (trou noir?) – apporte à ce problème un nouvel éclairage, tout comme aussi les tentatives statistiques qui évaluent, plus ou moins bien, à partir de considérations de l'évolution galactique, la fréquence des systèmes planétaires dans l'univers. Les exemples significatifs, mais en très petit nombre parce que proches, donnés par Van de Kamp, constituent l'argument solide de ce développement, en permettant à ces recherches plus conjecturales une base non contestable, et en les contraignant à rester raisonnables.

Des planètes, ici ou là, ailleurs qu'autour du Soleil, certes. Peter van de Kamp a montré quelques cas très sûrs, et cela reste fondamental. Et sur ces planètes – la Vie? C'est une toute autre question; et les probabilités qui interviennent dans bien des discussions classiques sur ce problème sont, le plus souvent, seulement de vagues intuitions. Peut-être la vie bien sûr... Et il n'est pas insensé de la chercher; les projets OZMA, SETI, etc. ont de beaux jours, derrière et devant eux. Nous verrons bien!

Une chose en tous cas est acquise.

Que ce soit en vue de l'exploration méthodique de l'univers proche et de ses propriétés dynamiques, ou que ce soit en vue d'y détecter des systèmes solaires frères du nôtre – une infinie patience est nécessaire. C'est cette patience qui est la vertu dominante des astrométristes, de ceux qui, comme van de Kamp, consacrent leur vie, et leur immense talent, à améliorer la précision des mesures et à en extraire quelques rares données – rares mais sûres. Cette patience qui exige une passion... Face à ces travaux, nous autres astrophysiciens, reprenons confiance; bien des découvertes nouvelles, pour étonnantes qu'elles soient, ne bouleversent pas la science; et le besoin reste aigu de cette précision exemplaire qui, au cours des siècles, a fait le bonheur et le prestige des astronomes.

JEAN-CLAUDE PECKER

CHAPTER 1

ASTROMETRY: HISTORICAL HIGHLIGHTS

The space-time relations, i.e. the motions of celestial objects are measured in a three-dimensional coordinate system, of which one coordinate is along, and the other two are perpendicular to the line of sight. The linear motion along the line of sight, the radial velocity, is obtained from the Doppler shift in the spectrum. The proper motion components, projected on the celestial sphere in the direction of right-ascension and declination are obtained by the technique and methods of astrometry. Their angular values may be reduced to linear measure if the distance is known.

(a) *Fundamental astronomy* deals with the positions of objects on the celestial sphere. The basic coordinates, right-ascension and declination are obtained from observations with transit or meridian circles, plus a sidereal clock. The observations are corrected for atmospheric refraction and instrumental errors; the resulting positions may be analyzed for changes in the equatorial coordinate system: precession and nutation, and for stellar proper motion and aberration. The attainable positional accuracy, about 0$''$.1, precludes useful measurements of parallax and orbital motion. Observations in fundamental astronomy have been carried out for some three centuries; the observing methods essentially have been visual. Gradually substantial improvements in accuracy have been obtained. The photographic technique has entered the field and has supplemented the visual approach through the use of wide-angle short-focus instruments, permitting coverage up to as much as one hundred square degrees of the celestial sphere on one photographic plate.

Since the beginning of the twentieth century *photography* with *long-focus* refractors and reflectors has resulted in large-scale portrayal of small portions of the sky, less than one degree in diameter; high accuracy of 0$''$.01 or better may thus be obtained for the position of a star on a background of distant stars. Long-focus photographic astrometry presently is the technique 'par excellence' for measuring stellar parallaxes, mass-ratios and perturbations. The same technique yields accurate values for the relative positions of binary-components.

(b) *Astrometry* has a long history, during which its various aspects were sequentially revealed in the order of the ease of discovery as determined by angular measure, by improved techniques and also by increased insight and interpretation of the observed phenomena. We shall briefly review some of the historical highlights.

The *precession* of the equinoxes – 50″ annually –, due to the gravitational attraction of Moon and Sun on the earth's equatorial bulge, was discovered by Hipparchus as early as 125 B.C.

The *heliocentric viewpoint* established by Nicolaus Copernicus in 1543 still accepted the planetary orbits as circular with the Sun off-center. The concept of *elliptical motion with the sun at focus, subject to the law of areas,* 'Keplerian motion', was introduced by Johannes Kepler in 1609, followed in 1619 by his *third* or *harmonic* law relating the orbits of the planets (known at that time) by the simple relation: cube of semi-axis major is proportional to the square of the period of revolution. Kepler's laws were derived from long series of observations by Tycho Brahe during the second half of the sixteenth century. The optical telescope was not invented till 1608; Kepler's laws rank among the greatest scientific discoveries of all time.

Kepler's three laws led to the universal law of gravitation established by Isaac Newton toward the end of the seventeenth century. This law in turn indicated small corrections to Kepler's laws; its universality also prepared for the later discovery of orbital motion in binary stars, and of perturbations.

True angular changes in stellar positions, or *proper motions* were not discovered until 1718 by Edmund Halley, from small displacements from positions recorded by Ptolemy in the second century A.D. Subsequent studies of tens of thousands stellar proper motions showed the latter to be constant in amount and direction, with exceptions of particular interest. Proper motions are now measured with high accuracy, often with probable errors well below 0.″001 annually; thousands of stars are known with annual proper motions exceeding 0.″5.

Stellar aberration due to the finite velocity of light (discovered by Olaus Römer in 1675) and the Earth's orbital motion around the Sun, was discovered in 1726 by James Bradley; its total amplitude range is nearly 41″. In 1728 Bradley found the *nutation,* a precessional term with a total amplitude of 23″ and a period of 18.6 yr, primarily due to the Moon.

(c) *Solar motion,* the Sun's motion relative to stars in our close galactic neighborhood was first established by William Herschel in 1783 from its secular parallactic effect in stellar proper motions. Later it was also revealed from stellar radial velocities. The observed total proper motion has two components, υ – toward the antapex of solar motion, and τ – perpendicular thereto. The υ-component approximates the *secular parallactic motion* and therefore, statistically, an analysis of the υ-components for a group of stars, gives us information about their *mean secular parallax*, which may be converted to *mean annual parallax* if we know the Sun's velocity from radial velocity observations. The τ-components together with statistical knowledge of radial velocities also give a statistical value for the average annual parallax of a group of stars. Both methods with appropriate knowledge of dispersion in distance yield values for the mean distance (van de Kamp, 1967).

The υ- and τ-component approaches have been successfully used in a number of studies not the least of which was the galactocentric revolution by Harlow Shapley referred to later, based on the statistical parallax value of only eleven Cepheid variables.

In 1802 Herschel noted the relative *orbital motion* for the two components of Castor and for other *binary stars*. The components of binary stars describe 'Keplerian' orbits around each other or, referred to a background of reference stars, each component orbits around their center of mass, which has a uniform rectilinear motion, as single stars have. The orbits of the two components around their center of mass differ 180° in phase, whereas their scales are inversely proportional to the masses of the respective components. Hence the two components move in non-rectilinear paths, which are the resultant of the uniform rectilinear motion of the mass-center and of the orbital motion of the individual components. Anticipating Chapters 11 and 13 we may say that for a binary star the orbital motion of each component may be considered a 'perturbation' of the proper motion of the (center of mass of the) system, caused by the gravitational action of the other component.

An accurate determination of *annual stellar parallax* was made in 1838 by Friedrich Wilhelm Bessel for the (double) star 61 Cygni, at that time the star of largest known proper motion (5".22 annually), hence judged to be nearby. Bessel employed the differential method ever since used, measuring the star on a background of reference stars within small angular separation from the 'parallax' star. The instrument used was the heliometer, a 'double image' micrometer, made by dividing the telescope objective in two halves which can slide by each other (Young, 1904). More accurate than the traditional micrometer and virtually free from possible sources of error (precession, refraction, instrumental) in 'absolute' or fundamental positions this differential method yielded an accurate value for the parallax of 61 Cygni; Bessel's value for the parallax was 0".35, now superseded by the 'modern' current value of 0".292.

Early photographic determinations of stellar parallaxes found their culmination in the long-focus photographic astrometric technique referred to earlier. The classical work by Frank Schlesinger, in the beginning of the twentieth century is the basis of all current work in this field (Chapters 2ff).

Currently parallaxes are measured photographically with high precision, down to $0''.002$ or better; hundreds of stars are known to have parallaxes exceeding $0''.1$ (Chapter 7).

In 1844 Bessel announced the existence of a *perturbation* in the proper motions of Sirius and of Procyon, based on visual observations with transit instruments over an interval of one century. The perturbations had amplitudes of several seconds of arc and therefore were well established. Bessel attributed both perturbations to the presence of invisible companions. These were seen later with large telescopes, the companion of Sirius by Alvan G. Clark in 1862 with the refractor of 46 cm aperture, which is now at the Dearborn observatory in Evanston, Illinois; the much fainter companion of Procyon was not seen until 1896 by J. M. Schaeberle with the refractor of 91 cm aperture at the Lick Observatory, Mount Hamilton, California.

In 1846 an unseen planet in our solar system, later named Neptune, was discovered from perturbations in the orbit of Uranus; in 1930 the distant planet Pluto was discovered in similar fashion.

Perturbations have been found in the orbits of binary and multiple stars; the well established classical cases are Zeta Cancri (Seeliger, 1888), Xi Ursae Majoris (Nörlund, 1905), and Zeta Aquarii (Strand, 1942). These three unseen companions, doubtless stellar, have not yet been seen; the unseen companion in the (by now quadruple) system Zeta Cancri, appears to be a white dwarf. Current photographic studies of perturbations, including subsequent sightings are reported in Chapters 13 and 14.

(d) An important historical statistical astrometric discovery was that of the *two star streams* by Jacobus Cornelius Kapteyn (1904). Reworded and reinterpreted as *ellipsoidal distribution* of stellar space motions (K. Schwarzschild, 1907, 1908), this phenomenon is frequently referred to as *preferential motion*, the greater mobility of stars along an axis pointing toward Sagittarius and toward Argo.

Meanwhile the *galactocentric revolution* (1914–1917) by Harlow Shapley (1918, 1919) paved the way for next steps. Radial velocities contributed to the analysis by Gustav Strömberg of the *asymmetry of stellar motions*, the increase in group motion with increasing dispersion in preferential motion for different groups of stars and other objects such as globular clusters (1925).

Both preferential motion and asymmetry were successfully explained by the theory of *galactic rotation* developed by Bertil Lindblad (1926, 1927). This in turn led to the mathematical formulation and discovery of differential galactic rotation by Jan Hendrik Oort (1927). Earlier (1922) Oort had made a study of *high velocity stars*, which also exhibit asymmetry.

The studies of interstellar absorption by R. J. Trumpler (1930) and others led to further improvement of our knowledge of the structure and dimensions of the Milky Way System. For more details see the author's reminiscent narrative (1965).

CHAPTER 2

LONG-FOCUS PHOTOGRAPHIC ASTROMETRY. TELESCOPE; MEASURING MACHINE

We shall now consider the precision branch of astrometry established in the beginning of the twentieth century, namely photography with long-focus telescopes portraying a small, virtually plane field of the celestial sphere. This technique introduced a marked advance in positional precision. We shall be primarily concerned with the astrometry of a central star, or stars, relative to a background of a small number of (distant) reference stars within the limited field of less than one degree. The techniques of observing measuring and calculating for this differential method were first developed by Frank Schlesinger with the long-focus visual refractor of Yerkes Observatory (focal length 19.37 m, aperture 102 cm) for the purpose of determining precise annual parallaxes (1910–11, 1924).

The high photographic accuracy in long-focus photographic astrometry results from several factors: long focal length, i.e. large scale portrayal, stability of photographic emulsions and precision measuring engines, and the relative or differential nature of the positional determinations. The attainable accuracy is limited primarily by telescopic and atmospheric effects which are kept small by maintaining stability of the optical parts, reducing the spectral band width, and observing one and the same star field always close to the same hour angle, preferably on meridian.

The early part of the twentieth century also witnessed the establishment of the multiple-exposure photographic technique by Ejnar Hertzsprung (1920)

TABLE 2.1
Atmospheric refraction and dispersion at altitude 45°

λ	R	Dispersion per 100 Å
4000 Å	61″.34	−0″.108
4500	60″.89	− ″.072
5000	60″.58	− ″.050
5500	60″.33	− ″.037
6000	60″.19	− ″.028
6500	60″.06	− ″.021
7000	59″.96	− ″.017
7500	59″.89	− ″.014
8000	59″.83	−0″.011

with the long-focus visual refractor of the Potsdam Observatory (focal length 12.5 m aperture 50 cm) for the purpose of determining accurate relative positions of well-separated components of visual binaries (Chapter 10)

(a) *Telescope*. The long-focus *refractor* has proven well suited for precise photographic measures. Proper choice of emulsion and filter keeps the range of light close to the wavelength corresponding to the minimum focal length of the focal curve (also called color-curve, achromatization-curve, or secondary spectrum) of the objective. The photographic position still depends on the residual energy distribution of the star's spectrum, as filtered by the objective (transparency and focal curve), filter, and emulsion. The effective wavelengths of the star images depend on the spectrum and, to some extent, on the magnitude; however, with proper choice of filter and emulsion, this dependence may be reduced to a minimum.

The rapid decrease in atmospheric *dispersion* (Table 2.1) toward longer wave lengths proves advantageous also to the photographic technique as applied to visual refractors, even though these originally were not planned for photographic use. As an illustration, the Sproul visual refractor with an achromatic crown-flint glass objective (aperture 61cm, focal length 10.93m, scale 1mm = 18″.87) is used in conjunction with 13 × 18 cm Eastman Kodak 103aG plates and a 'minus blue', at present Schott OG-515, filter which practically eliminates all radiation on the blue side of approximately λ5100. An effective bandwidth of about 600 Å is obtained centered on wavelength

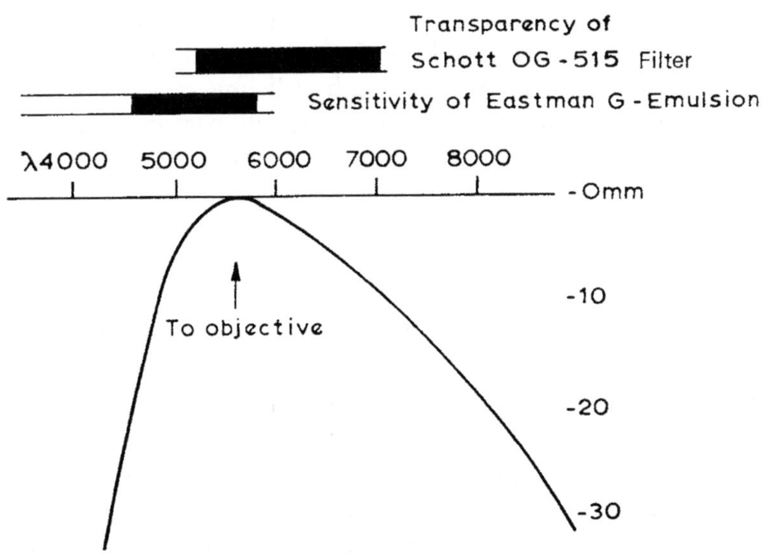

Fig. 2.1. Focal curve of Sproul 61 cm objective.

$\lambda 5607$ at minimum focal length. The *photovisual* technique yields sharp images with effective wavelengths ranging from about $\lambda 5480$ for spectral type A0 to $\lambda 5525$ for spectral type M0. The corresponding difference in refraction constant (altitude 45°) is 0ʺ016, the A-star being comparatively that much closer to the zenith than the M-star (van de Kamp, 1967; Wanner, 1968).

The long-focus refractors have a *focal-ratio* (focal length divided by aperture) of something like $f = 15$. A typical scale value is $1\mu = 0''02$. The diameters of a well-defined sharp, well-blackened star image range from about 0.05 mm to 0.1 mm (about 1ʺ to 2ʺ). Sources of error in the differential method are differences in magnitude and color between central and reference stars. Limitations in positional accuracy are also due to optical field errors, which may vary with temperature. The principal off-axis aberration of many existing long-focus objectives is *coma*, which results both in scale and magnitude effects in the image plane (Strong, 1958). Coma is proportional to $1/f^2$ and to the angular distance from the optical axis. Hence both reduction of measured field, and if desired and practical, reduction of telescope aperture lead to increased accuracy (Ianna, 1965). Other possible sources of error are maladjustment of focus and collimation, and *spherical aberration* resulting from different focus for different zones. Proper ventilation of dome and telescope is essential (Lippincott, 1965). Errors due to the above effects: spherical and other aberrations, coma and dispersion are minimal in long-focus photographic astrometry by its differential nature. Only changes in these effects introduce errors.

According to Rayleigh an optical instrument may be considered perfect if the difference between the longest and the shortest optical paths at the focus does not exceed one quarter of a wavelength. All images – photovisual or photographic – obtained within $4f^2\lambda$ of a theoretically perfect optical system are equally good; λ is the wavelength that corresponds to minimum focal length. Generally for long-focus refractors *Rayleigh's criterion* is between 0.5 and 0.7 mm; for the Sproul refractor at full aperture ($f = 18$, $\lambda = 5607$), it is 0.7 mm.

(b) Early parallax work was done mostly by refractors; reflectors were at a disadvantage because of very limited field and changes in figure. The situation has gradually changed, from glass to pyrex to quartz reflectors culminating in the successful operation and results obtained with the long-focus *reflector* of 155 cm aperture of the United States Naval Observatory (USNO) (Strand, 1964, 1971). This fine instrument designed for and at present dedicated to photographic astrometry saw first light in March 1964; it is situated at the Observatory's remote observing station in Flagstaff, Arizona. The telescope has a parabolic fused silica primary, with a flat secondary of 0.9 m diameter reflecting the light back to a Cassegrainian focus (although this is not a Cassegrain reflector). The secondary support system is designed in such a way that the secondary is always perpendicular to the optical axis as the telescope sags in different ways with different settings. The scale of the instrument is $13''.55\ mm^{-1}$, corresponding to a focal length of 15.22 m. The coma-free field is approximately 25′ in diameter, which fits nicely on a standard 13 × 18 cm plate. Designed to observe stars in the magnitude range 12 to 18, there is at present no provision for magnitude compensation although filters with neutral-density occulting spots will shortly be added to the system. Other astrometric reflectors (Fan Mountain, Pino Torinese) have been built and are now in operation.

(c) *Measuring machines.* In the early days of photographic parallax determinations, simple one-screw measuring machines, such as described elsewhere (van de Kamp, 1967) were used to obtain the relative positions of images on the photographic plate. With increased accumulations of observational material and the desire and possibility for greater measuring accuracy, the last two decades have witnessed the construction of improved measuring machines with a marked reduction in measuring error and a reduction if not complete elimination of systematic errors resulting from human judgment in bisecting a star image. The present state of the art will be illustrated by two reports on measuring machines which have been in active operation for a decade or more.

First a brief description by Dr Robert S. Harrington of the measuring machine(s) now in use at the *United States Naval Observatory* in Washington D.C. The plates are measured on a specially built measuring machine designed as part of the total astrometric system. The machine automatically and impersonally centers on each image, based on two orthogonal integrated photoelectric scans. The heavy granite stages move on air bearings, and the positions of the stages are recorded to a micron through two Moiré fringe 'picket fence' encoders, consisting of two long pieces of glass (gratings) on which very many thin lines have been accurately ruled crosswise. These are placed side by side, one on a fixed stage and one on a movable one. A beam of light shines through the system onto a photodetector. As one stage moves, its grating alternately interferes and coincides with the fixed one, producting an oscillating intensity on the detector. Counting the number of oscillating cycles, and knowing the grid spacing of the gratings, yields the distance that one stage has moved with respect to the other.

The formal local internal accuracy of the machine (generally known as SAMM, for Strand automatic measuring machine), is around 0.7 microns.

With this machine, it is possible to stay even with the output of the astrometric reflector with the normal level of activity, meaning that no backlog of unmeasured plates can accumulate.

The USNO has a second measuring machine (known as Starscan, or just SCAN), which was built much later than, but in many ways patterned after SAMM. SCAN was designed to measure larger astrographic plates rather than long-focus plates and has several features with this in mind. It can handle a 25 × 25 cm field, is of somewhat higher precision (0.4μ), and, in the measuring mode, operates approximately a factor of seven faster than SAMM. The machine is under the control of a mini-computer, and it will shortly be upgraded to include a mode of operation whereby it can scan an entire plate and locate and measure every image on it. Besides recording orthogonal coordinates, the machine also gives a measure of the integrated density of the image, making a form of photographic photometry possible with suitable careful calibration.

(d) The following detailed description of the *Grant* measuring machine at the *Sproul Observatory* has been prepared by John L. Hershey. The urgent need for faster measuring techniques in the mid-1960's prompted planning with other observatories and the Grant Instrument Company of Berkeley, California, for the design and construction of a two-screw measuring machine satisfying the Sproul specifications. Four machines were made simultaneously with a number of options to satisfy the needs of each prospective customer. At that time the one-coordinate Grant spectrum comparator had proven successful in spectroscopy at numerous institutions. At present two of the two-coordinate Grant machines are in use for long-focus astrometry; the Sproul machine has been in daily use since its installation in 1971 and more than 60 000 plates have been measured.

The primary desideratum for a plate measuring machine in differential astrometry is the most accurate determination of changes from plate to plate of the relative positions of star images in two coordinates. Formal probable errors of parallaxes from plate series extending over less than ten years have been reaching below $0\overset{''}{.}002$ which is 0.1μ or 1000 Å in linear measure on the Sproul plates. The differential nature of long-focus astrometry contributes to this high accuracy.

The Grant machine has capacity for 25 cm square plates; the Sproul plates are 13 × 18 cm, and often only a 5–8 cm range is used. The use of a small fractional range of capacity also contributes to accuracy of the results.

The tables of the machine ride on steel ways of rectangular cross section. The detection of position is carried out by rotary pulse generators which detect the rotation of the precision screws which drive the table along the ways. The pulse generators give 2000 counts per milimeter which is one rotation of the screw. The counts are displayed visually and are recorded by a card punch. The measurement of a star image position is accomplished, as in the one-coordinate Grant machine, by superimposing an image density profile upon its mirror image on an oscilloscope screen. The profiles are superimposed on separate screens for each coordinate, under manual control of the screws by handwheels. By judicious settings of the scanning slit heights, width, and the oscilloscope electronics, well peaked profiles of the circular images can be formed for precise settings.

The 60000 plates measured on the Grant machine during the past decade yield many comparisons with measurements of plate series by visual bisection with cross wires on the old 'classical' one screw machines. Usually a decrease of 30% in the error of reduced star positions is achieved with the Grant machine, corresponding to a doubling of plate weight. In some cases the error has been reduced to one half the value from manual measuring of the identical plates and identical reference stars. This accuracy is achieved with two settings on the central star and one setting per reference star image of the Grant profiles as compared with four settings of visual bisection with cross-wires. The high accuracy of the Grant machine could conceivably be approached by a large number of visual cross-wire bisections, by an expert measurer with intense concentration. However, the Grant machine's accuracy can be achieved by an unskilled operator with much greater ease of operation and at much higher speed, and for longer time intervals of measuring. Moreover the Grant measurements are far less personal than measurements obtained by visual bisection. The Grant machine makes the greatest gain in accuracy on the large images from brighter stars or poor seeing. Although some personal judgment is required in overlapping asymmetric profiles, any personal differences are independent of the direction of approach.

Most of the plates have been measured with only one setting on reference star images and two settings on the central star which opens and closes the measuring sequence. The typical error of repeatability of the position of one image after plate reduction in a stellar reference frame is approximately one micron which is less than the intrinsic error of the photographic process. Experiments with several plate series have been made with two independent measurements of each plate. By comparing the reduced positions from two measurements on each of several hundred plates the error of one measurement of the Grant machine has been found to be approximately one-half the error of the photographic positional information of images in a stellar reference frame. Thus the maximum improvement possible in the present photographic technique by a perfect measuring machine should be only about 10%. This limit would predict a 5% decrease in probable error when the mean of two independent measurements is taken. Reductions in error of 5% or less have been found experimentally where parallax solutions have been made on several plate series from single measurements and from the mean of two measurements. It is a matter of personal opinion whether it is wiser to double plate processing time or increase the number of plates by 10% or less to achieve an equivalent level of accuracy.

Although the setting on images is done manually on the Grant machine, the operators have found that the process of superimposing the profiles in both coordinates, in an iterative fashion, soon becomes a matter of reflex action and is accomplished in a matter of seconds with no physiological strain. The manual mode of positioning allows the operator to avoid errors from obvious image defects, such as adjacent dust specks, which an automated bisection device would not recognize. An automatic centering system would center on an image in a few seconds less; however an appreciable fraction of the measuring time is spent in moving from star to star, inserting and removing plates, and recording plate data and date of measuring. Hence an automatic centering system would not speed the entire measuring process by a large factor when only 20 to 50 images are measured per plate.

The long term stability of the machine has been monitored by measuring test plates twice a month since 1971. No long term changes are apparent in these measurements.

Thermal effects can be a problem in a measuring machine when a long interval of time is required to measure a plate. The Sproul plates are usually measured in less than ten minutes which is short compared with expected thermal changes of mechanical or electronic parts in a room with controlled temperature.

The counters are zeroed at the beginning of a plate measurement on a central image. At the end of the plate measurement the central images are again measured. The operator can then see any change larger than measuring error due to failure of the counters or any other cause.

The Grant machine is operated daily, at somewhat under 8 hours per day, less in the summer, but plates are being measured at approximately the same rate that photographs are taken at the telescope. Compared with the old one screw machines, this rate represents a factor of five or more in the speed of the process of measuring and reducing plates, even though the average number of reference stars measured has increased somewhat. The increase in speed is due in part to the use of the computer for plate reductions from the punched measurement cards. See also Hershey (1975b).

CHAPTER 3

OBSERVATIONAL ERRORS. INSTRUMENTAL EQUATION

(a) *Accidental errors.* With the Sproul telescope the differential method i.e. central relative to reference stars, yields a precision of about 1μ or $0\overset{''}{.}02$ for the average of two exposures. By taking up to five exposures per plate and up to four plates on any one night a precision of 0.5μ or $0\overset{''}{.}01$ may be reached. And by combining several nights within one year into a normal point an accuracy of 0.3μ or $0\overset{''}{.}005$ – by a supreme observational effort as low as 0.1μ or $0\overset{''}{.}002$ – may be approached. It appears difficult if not impossible to reduce the error any further; the law of diminishing returns indicates the existence of a 'year error' of about 0.1μ or $0\overset{''}{.}002$ (Chapter 5).

The multiple exposure method for double stars (Chapter 10) makes use, if need be, of a grating for reducing magnitude error, and a large number of exposures (up to forty or more) are taken. A yearly normal point based on six or more multiple-exposure plates yields a precision of about 0.1μ or $0\overset{''}{.}002$, the same as may be approched in the differential method of central star measured on background of reference stars.

Although high accuracy may thus be obtained in terms of accidental errors, there remains the serious possibility of systematic errors in the differential approach, particularly for series of observations extending over long time intervals and therefore of particular concern in studies of perturbations (Chapter 13ff). And, of course there is the possibility of annual systematic errors which could affect parallax determinations.

(b) *Systematic errors.* We start with the classical statement by Kapteyn (1922): "I know of no more depressing thing in the whole domain of Astronomy, than to pass from the consideration of the accidental errors of our star places to that of their systematic errors." Kapteyn referred to fundamental star positions, covering more than two centuries of visual observations with meridian circles. Illustrations of this matter are the Catalogues of L. Boss (1910) and of B. Boss (1937), in which systematic corrections and weights of star positions are listed for different instruments at different observatories. The errors are primarily considered functions of right-ascension and declination, but also of apparent magnitude. Leaders in the studies of systematic errors in fundamental star places have been Lewis Boss, A. Auwers, A. Kopff, and currently Walter Fricke.

The concern and studies of these astronomers should alert us to the likelihood of a comparable situation existing in the technique of long-focus photographic astrometry. Schlesinger drew attention to the more obvious sources of error, such as depend on hour-angle. Early parallax determinations covered only a few years of observation. The study of systematic errors and their possible changes over long time intervals has only started to play a role comparatively recently, when observational evidence has revealed their occurrence. Kapteyn's concern is found to hold equally for the current long-focus photographic technique. Some refractors have been in use for over six decades, and there is evidence of systematic changes over long time intervals, corresponding with events in the history of the objective.

Precise, quantitive information on systematic errors is slowly emerging. We give a brief report of what appears to be known for the long-range series of photographs obtained at the Sproul Obervatory. These observations, often extending over more than four decades, were planned and are a *sine qua non* for long-range studies of orbital motion and of systematic errors. The series may be subject to temporal changes (gradual or abrupt) in photographic portrayal, due to a variety of causes, such as changes in the optical performance of the objective, changes in filter and in photographic emulsion.

(c) *Instrumental equation: Sproul refractor*. To separate possible systematic changes in astrometric portrayal from true cosmic effects, we realize that a stellar path implies a continuous time displacement pattern and does not suffer sudden changes. On the other hand abrupt instrumental changes over long time intervals are revealed as discontinuous effects, corresponding most likely, as experience has demonstrated, to instrumental adjustments. The general procedure should be to detect and allow for such adjustments or *instrumental equations*, that will provide continuity with earlier and later material. Obviously the same intensive coverage aimed at studying parallax, proper motion and orbital motion also serves to establish and allow for any temporally limited instrumental equation, recognized as an abrupt discrete deviation, a discontinuity from the smooth stellar path.

How constant has been the performance of the Sproul refractor since intensive observations were started in 1937? There have been several critical dates in the history of the Sproul objective, which was removed from its cell in 1941.82 to permit adjustment, carried out by Carl Lundin, Jr. of the Warner and Swasey Company. In 1949.21 the old aluminum cell was replaced by a snug-fitting cast iron cell, constructed by the Perkin-Elmer Company, and the objective was installed by Kurt Opperman of that company. An adjustment for collimation was made in 1957, and in 1966 the objective was cleaned and replaced, again by Kurt Opperman, in conjunction with the general renovation of the mounting and mechanical operation of the refractor by the Wilmot Fleming Company (Wanner, 1968).

Thus far the most important discontinuity appears to be an instrumental equation, related to difference in color between central and reference stars, primarily in right ascension over the interval 1941.82–1949.21 (van de Kamp, 1977a). The post-1949.21, as well as the pre-1941.82 portrayal, are adopted as 'standard'. The existence of the systematic 'break' in right ascension in 1941 was recognized shortly afterward and has been reported several times (Lippincott, 1957, 1971). The 1949 'break' first was the subject of thorough astrometric study of one particular starfield, for which Hershey (1973a) found a correction of $+2.9\mu$ over the interval 1942–1948 for the red star AC+65° 6955. Other extreme examples of this instrumental equation have been established for Barnard's star (van de Kamp, 1975, 1977a) and VV Cephei (van de Kamp, 1977b), both very red stars. For the former an instrumental correction of $+2.8\mu$, for the latter $+2.4\mu$ is found in RA, over the interval 1942–1948. A correction of $+2.0\mu$ has been found for the red stars BD+43° 4305, (van de Kamp and Lippincott, 1980), Groombridge 1618 and for Wolf 294; $+1.0\mu$ for Ross 128 (Hershey *et al.*, 1980). These equations amount therefore to as high as 0".05 for the reddest stars. On the other hand, for van Maanen's star (Hershey, 1978), where there is no sensible difference in color between central star and reference background, no instrumental equation appears to exist. Also, Epsilon Aurigae is bluer than the reference background; a negative instrumental correction in RA over the critical interval is expected and is indicated by the residuals from the parallax solution; they point to a correction of -1.0μ over the interval 1942–1948 (van de Kamp, 1978a). Over the same critical interval much smaller instrumental corrections in declination are indicated for certain red stars, e.g., -1.0μ for AC + 65° 6955, -0.5μ for VV Cephei and for BD + 43° 4305 (possibly -0.5μ for Barnard's star).

This instrumental equation being dependent on color is a matter of concern for the many red stars on the observing program, measured on a background of non-red stars.

However the interval 1942–1949 in RA has now become a relatively minor segment in the perspective of the long-range – up to 60 years – Sproul plate series. The strongest coverage for most stars is in the 1960–1980 interval. In no case is the 1942–1949 interval critical to orbital interpretation. In most cases it can be included or omitted with very little effect.

Measurements of series of plates on some two hundred different starfields show that the optical performance of the Sproul refractor was not affected by appreciable instrumental equations in 1957 or 1966; formal values far below 0.5μ are indicated. Since 1949 the astrometric portrayal has been quite stable, considering the size of 1μ, relative to image and plate size, size of telescope and length of time coverage (Lippincott, 1978; Hershey *et al.*, 1980). There is evidence for yearly systematic errors, perhaps due to seasonal atmospheric and instrumental effects, changing from year to year.

A preliminary astrometric history of the McCormick refractor is given by Probst (1977). The astrometric reliability of the Allegheny long-focus refractor has been discussed by Kamper (1971); a relative rotation of 35° was found over the interval 1944 to 1969 between the two optical components.

We may well have to reckon with the possibility that the optical behavior of any one telescope differs from star to star, i.e., depends on RA and Decl. (assuming the precautionary measure of always observing close to the meridian). How astrometric portrayal compares for different telescopes is another matter; there is evidence, *a priori* not unexpected, for instrumental equations between different telescopes. This is not a serious matter in differential astrometry, as long as the optical behavior, i.e., astrometric portrayal, by any one telescope for any starfield remains constant.

(d) *Time of night effect.* Another small, systematic effect has become evident, namely a dependence of declination positions on the time of the night or what for any one star amounts to the same, time of the year. This effect appears in an analysis of Sproul astrometric accuracy. Some 9511 night residuals from 124 Sproul parallax determinations indicate a positive systematic error of 0.2μ or $0\overset{''}{.}004$ in declination between $22\overset{h}{.}5$ and $1\overset{h}{.}0$ EST relative to earlier and later times of the night (Lippincott, Table XI, 1971). A later intensive homogeneous series of observations of 758 night residuals, over the interval 1950–78 for Barnard's star, gives the following results for the required correction to eliminate the time of night effect in declination:

$> 1\overset{h}{.}5$ EST	$+\,0.10\mu$
22^h–$1\overset{h}{.}5$ EST	$-\,0.34\mu$
$< 22^h$ EST	$+\,0.16\mu$

i.e., a differential correction of -0.4μ to -0.5μ for the 'middle' of the night relative to the 'ends'.

The source of this systematic behaviour in declination for the Sproul telescope is still not explained. There appears to be no correlation with hour-angle (which has very little range because of the Sproul policy of hour-angle restriction). A slight correlation with temperature is indicated; no dependence on barometric pressure is found. Clearly this error affects parallax determinations based on declination measures only, and generally would lead to a slight increase in the parallax values from declination observations made in the northern hemisphere. Apparently the declination effect does not cancel for a star measured on a background of reference stars; i.e., a parameter of the central star appears involved. Whether this parameter is color and/or location in the sky, or something else, is not clear at the moment. However, for a large number of parallax determinations at the Sproul Observatory on the average the declination results for parallax are only 0.″0018 ±0.″0011 larger than the right ascension results, perhaps a reason for some guarded satisfaction (Lippincott, 1971).

CHAPTER 4

STELLAR PATHS. REDUCTION OF MEASUREMENTS

The principal purpose of the astrometric study of the paths of single stars is to determine proper motion and parallax, (Chapter 5) and, in the long run quadratic time effects related to perspective acceleration (Chapter 8), as well as possible perturbations due to unseen companion objects (Chapter 13ff). Dynamical acceleration effects, due to galactic rotation or other galactic influences, are far too small to be of any concern at present. For established double stars there are the additional orbital motions around the center of mass, which is assumed to have the property of a single star (Chapter 11).

All these astrometric studies are based on the differentially observed path of the star or stars relative to a background of reference stars about magnitude 10 or fainter. In conventional parallax studies the path may include only a few dozen plates spread over an interval of a few years. In studies involving orbital motion, perturbation and/or quadratic time effects, the observed path may be based on hundreds of plates taken during an appreciable fraction of a century. The observation and reduction methods are essentially the same in all these studies and are now considered. For further details see Principles of Astrometry, Chapter 5ff (van de Kamp, 1967).

(a) *Image plane and tangential plane. Equatorial and standard coordinates.* In long-focus photographic astrometry we are concerned with the portrayal of a small portion of the celestial sphere on an *image plane* registered by a photographic plate, which corresponds, via the optical axis of the system, to a *tangential plane* on the celestial sphere. The unit of length is the radius of the celestial sphere or that of its image, the focal length. A point S_0 on the celestial sphere corresponds to point S on the tangential, or on the image plane, which is assumed to be parallel to the tangential plane. Referred to the tangential point T we have the rectangular coordinates x and y, where the y-axis is tangent to the hour circle at T, positive in the direction in which the distance of T from the north celestial pole is $< 180°$. The x-axis is perpendicular to the y-axis, and is positive in the direction of increasing right ascension.

The *standard coordinates* x and y are related to the *equatorial coordinates*

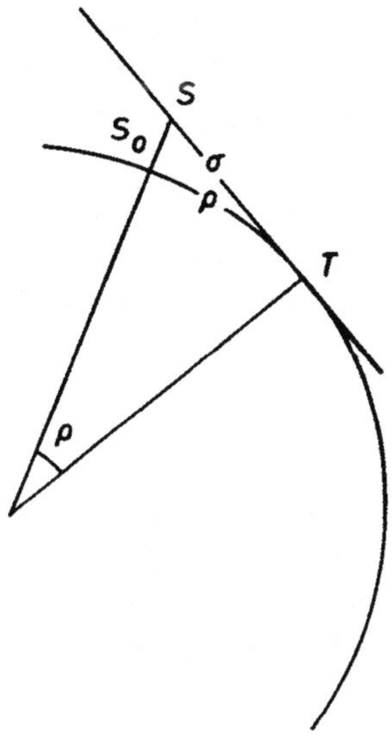

Fig.4.1. Plane coordinates. Cross section of sphere and plane. T = tangential point; S = central projection on plane of star S_0 on sphere.

α and δ as follows:

$$x = \frac{\cos\delta \sin\Delta\alpha}{\sin\delta\sin\delta_0 + \cos\delta\cos\delta_0\cos\Delta\alpha},$$
$$y = \frac{\sin\delta\cos\delta_0 - \cos\delta\sin\delta_0\cos\Delta\alpha}{\sin\delta\sin\delta_0 + \cos\delta\cos\delta_0\cos\Delta\alpha}, \tag{4.1}$$

and, vice versa:

$$\sin\Delta\alpha = \frac{x\tan\delta}{\sin\delta_0 + y\cos\delta_0},$$
$$\sin\delta = \frac{\sin\delta_0 + y\cos\delta_0}{\sqrt{1 + x^2 + y^2}}. \tag{4.2}$$

Here α_0, δ_0 refer to the tangential point, α as δ to the object S, while $\alpha = \alpha_0 + \Delta\alpha$ and $\delta = \delta_0 + \Delta\delta$.

For a small angular field we may write:

$$x = \Delta\alpha\cos\delta_0 - \Delta\alpha\Delta\delta\sin\delta_0,$$
$$y = \Delta\delta + \tfrac{1}{4}(\Delta\alpha)^2\sin 2\delta_0, \tag{4.3}$$

and

$$\Delta\alpha = x\sec\delta_0 + xy\sec\delta_0\tan\delta_0,$$
$$\Delta\delta = y - (x^2/2)\tan\delta_0. \tag{4.4}$$

Formulae (4.3) and (4.4) are useful in dealing with the positions of asteroids and comets where transformation from standard to equatorial coordinates, since the object moves rapidly from field to field, and spherical astrometry is required.

For studies of a *stellar* path covering one and the same field over an interval of a good fraction of one century, x and/or y are well below 0.002, and within a negligible error the above relations become simply:

$$x = \Delta\alpha \cos \delta_0, \quad y = \Delta\delta$$
$$\Delta\alpha = x \sec \delta_0, \quad \Delta\delta = y. \tag{4.5}$$

For an extended stellar path, an increasing disparity between rectangular standard and spherical coordinates exists, which may be represented to a high approximation as a slowly changing orientation of the equatorial with respect to the rectangular coordinate system, amounting to

$$\Delta\theta = \Delta\alpha \sin \delta, \tag{4.6}$$

where $\Delta\alpha$ is the projection of the path in the x-coordinate measured in right ascension, and δ is the declination. In comparing results expressed in rectangular and in equatorial coordinates, the above change in position angle must be kept in mind, and secular effects should be considered, especially for stars of large proper motion situated at high declination; the effect on the components of the annual parallactic motion is negligible.

The difference between spherical and plane portrayal of a stellar path 1000″ long remains below 0.″01, and is therefore generally negligible.

(b) *Scale, Orientation, and tilt effects.* The principal geometric effects on the measured positions are changes in scale and orientation resulting from the following causes:

Observational

Instrumental (telescope, measuring machine)
Spherical (refraction, aberration)

Cosmic

Proper motions of reference stars.

Variations in scale due to changes in temperature and barometric pressure are negligible for the present purpose. The linear scale effect due to astronomical refraction is 0.00029, and is incorporated in the scale value of the telescope. The linear scale effect due to aberration is at most 0.0002, and its possible effect on parallax is negligible. Because of the small angular extent of long-focus plates, small changes in refraction and aberration may be

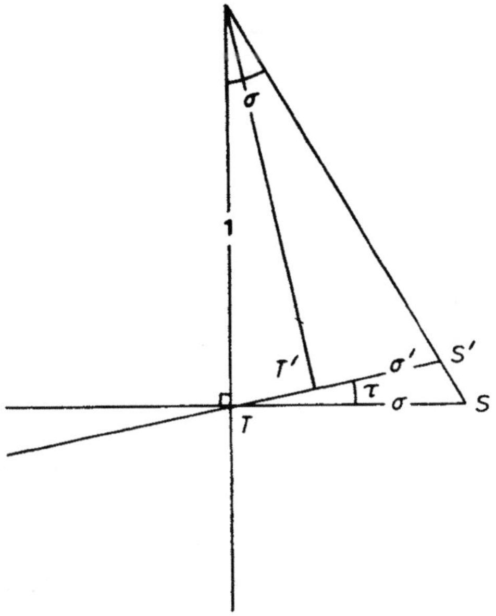

Fig. 4.2. Effect of plate tilt τ on measured position S', compared with ideal position S.

considered to be linear functions of the position of the star on the plate, and the nonlinear terms may be ignored.

In general a linear transformation or reduction is all that is required to take care of the observational geometric effects; the cosmic effects require special treatment and will be described in Chapter 6.

Ideally the photographic plate coincides with the image plane. If not, we introduce the base or optical center T' of the plate, i.e., the point at which a perpendicular from the optical center falls on the plate. The point S on the ideal image plate is now portrayed as a point S' on the actual photographic plate. A simple calculation shows that the following correction is required to observed (measured) standard coordinates:

$$\begin{aligned} &\text{in } x \quad px^2 + qxy \\ &\text{in } y \quad pxy + qy^2 \end{aligned} \tag{4.7}$$

where p and q are the standard coordinates of T' with respect to T.

The effect of *plate tilt* is generally negligible for long-focus instruments. For an extreme x or y of 20′ and values of 1′ for p and q the quadratic terms in the above expression would be only 0″.002.

(c) *Reference Stars; Standard frame, linear plate constants.* The positional measurements of a stellar path are made on a background of several faint

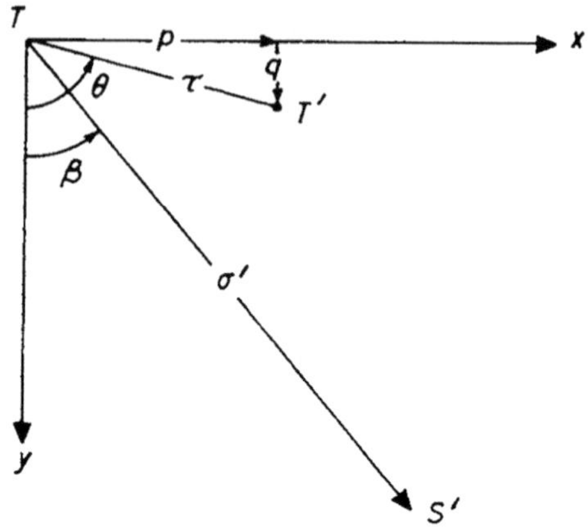

Fig. 4.3. Relation between position S' and optical center T' of plate.

stars, about magnitude 10 or fainter. Ideally all reference stars should have the same brightness and spectrum (color) as the central star; in practice this can only be approximated. The brightness of the central star, if necessary, is reduced by a rotating sector to obtain approximate magnitude compensation between central and reference stars; grating techniques may also be used. The photographic plates yield tangential, or *standard*, coordinates in the image plane that are closely oriented with the directions of right ascension and declination. When comparing the different plates of a series of observations of the same field, all that is necessary is to reduce the measurements to a common origin, scale, and orientation by a linear transformation. This is provided by a set of n reference stars. Three reference stars are a minimum requirement; generally four to six reference stars are used, rarely more.

Let X', Y' and x', y' be the measured coordinates of central and reference stars as obtained at the measuring machine. In order to permit a comparison of measured positions on different plates, a reduction is made to a *standard frame* based on the reference stars. Its coordinates are admittedly and intentionally approximate: they may be taken as the positions of the reference stars referred to the equator of any convenient epoch (2000 is now commonly used) and rounded off to say 0.01 mm. The standard frame is simply an idealized or fictitious standard plate to whose origin, scale and orientation all other plates in the series are reduced.

The coordinates of the standard frame will be denoted by the subscript s; the positions x_s and y_s, defining the standard frame of reference, are relative to their mean position, i.e.,

$$[x_s] = [y_s] = 0. \tag{4.8}$$

All measured positions can now be reduced to the scale, orientation, and origin of the reference frame (x_s, y_s) through plate constants *a*, *b*, and *c*, which are given by the linear equations of condition for each reference star;

$$\begin{aligned} a_x x_s + b_x y_s + c_x &= x_s - x', \\ a_y x_s + b_y y_s + c_y &= y_s - y'. \end{aligned} \tag{4.9}$$

A least squares solution gives

$$\begin{aligned} a_x &= \frac{[y_s^2]\,[x_s(x_s - x')] - [x_s y_s]\,[y_s(x_s - x')]}{[x_s^2]\,[y_s^2] - [x_s y_s]^2}, \\ b_x &= \frac{[x_s^2]\,[y_s(x_s - x')] - [x_s y_s]\,[x_s(x_s - x')]}{[x_s^2]\,[y_s^2] - [x_s y_s]^2}, \\ c_x &= -\frac{[x']}{n}, \end{aligned} \tag{4.10}$$

and similar expressions for a_y, b_y, and c_y.

For the central star the position X, Y reduced to the standard frame is given by

$$\begin{aligned} X &= X' + a_x X_0 + b_x Y_0 + c_x, \\ Y &= Y' + a_y X_0 + b_y Y_0 + c_y, \end{aligned} \tag{4.11}$$

where X_0 and Y_0 are adopted values of X and Y. Thus the position X_0, Y_0 is *rigorously* corrected for plate constants, but the values of X and Y obtained through the relation (4.11) remain uncorrected to the extent of

$$\begin{aligned} &a_x(X - X_0) + b_x(Y - Y_0) \text{ in } x, \\ &a_y(X - X_0) + b_y(Y - Y_0) \text{ in } y. \end{aligned} \tag{4.12}$$

(d) *Dependences; geometric accuracy.* Schlesinger has shown that, for linear plate constants, insight is gained by expressing the reduced position as an explicit linear function of the measured coordinates. The necessary transformation is obtained by substituting Equations (4.10) in Equations (4.11), which leads to

$$\begin{aligned} X = X' + \Bigg\{ &\frac{X_0\{x_s[y_s^2] - y_s[x_s y_s]\} + Y_0\{y_s[x_s^2] - x_s[x_s y_s]\}}{[x_s^2]\,[y_s^2] - [x_s y_s]^2} \times \\ &\times (x_s - x')\Bigg\} - \frac{[x']}{n}, \\ Y = Y' + \Bigg\{ &\frac{X_0\{x_s[y_s^2] - y_s[x_s y_s]\} + Y_0\{y_s[x_s^2] - x_s[x_s y_s]\}}{[x_s^2]\,[y_s^2] - [x_s y_s]^2} \times \\ &\times (y_s - y')\Bigg\} - \frac{[y']}{n}. \end{aligned} \tag{4.13}$$

The resulting reduction statement, regardless of the zero point of the measured coordinates, therefore is

$$X = X' + [D_i(x_s - x')_i], \qquad Y = Y' + [D_i(y_s - y')_i]. \tag{4.14}$$

where $i = 1, \ldots, n$.

The quantities

$$D_i = \frac{X_0\{x_{si}[y_s^2] - y_{si}[x_s y_s]\} + Y_0\{y_{si}[x_s^2] - x_{si}[x_s y_s]\}}{[x_s^2][y_s^2] - [x_s y_s]^2} + \frac{1}{n}, \tag{4.15}$$

are named *dependences*. It is obvious that $[D] = 1$, and, because of the least-squares procedure, that $[D^2]$ is a minimum. In the plate-constant method, X and Y are *implicit* functions of (x') and (y'); the dependence method provides an *explicit* expression.

The dependence reduction (4.14) may be written as follows:

$$X = [Dx_s] + X' - [Dx'], \qquad Y = [Dy_s] + Y' - [Dy']. \tag{4.16}$$

The position $[Dx_s]$, $[Dy_s]$ defines a point close to the central star, which is rigorously corrected for plate constants and is called the *dependence center*; $[Dx']$, $[Dy']$ is the measured *dependence background*. The dependences are the barycentric coordinates of the central star with respect to the reference stars.

Plate reduction may now be accomplished by using the dependences with commonly are expressed to three decimals. The dependence method leaves only a small segment uncorrected, the so-called *plate solution*, sometimes called *offset*:

$$\xi = X' - [Dx'], \qquad \eta = Y' - [Dy'], \tag{4.17}$$

which remains uncorrected to the small amount (4.12). Theoretically, the plate solutions should vanish for the position X_0, Y_0, but, because of the limited number of decimals in D, small plate solutions exist for this position. We now have

$$X = [Dx_s] + \xi; \qquad Y = [Dy_s] + \eta. \tag{4.18}$$

The zero point of measurement is eliminated for ξ, η, while x_s, y_s refer to their mean (4.8).

In the early, pre-computer days, for short-range parallax series one and the same set of dependences was used as a quick, efficient and explicit way for calculating ξ and η (formula (4.17)). Careful orientation at the measuring machine provided rigorous plate-constant reduction with one set of dependences for a large number of measurements for plate solutions ξ, η as large as 0.5 mm. Gradually for long-range series, involving orbital motion or perturbations, stars of appreciable proper motion required the use of successive sets of dependences and a shift from ξ, η (4.17) to X, Y (4.18) became mandatory.

The dependence reduction method no longer has any computational advantage. But the dependences, and their changes with time, retain the important significance of representing the contribution or *weight of the position of each reference star* to their weighted mean background. No study of a stellar path can be intelligently planned without the concept and knowledge of the dependences, and also their change with time. In Chapters 6 and 8 we shall see further examples of the usefulness and elegance of the concept of dependences.

An obvious use of dependences is the evaluation of the *geometric accuracy* of the reduced position of the central star on the background of reference stars. The 'geometric' error of the reduced position is proportional to $(1 + [D^2])^{1/2}$, assuming the same inherent positional accuracy for central and reference stars. The error of the reduced position is primarily due to the error of the central star which enters with the full amount. For the case of the latter being located at the geometrical center of the reference stars, resulting in equal dependences, an increase of the number of reference stars from four to eight up to twenty leads to a reduction in the geometric error of the reduced position of only six percent and ten percent respectively. Most likely the actual decrease in error would be less because of greater difficulty, with increasing number to find suitable reference stars in terms of magnitude and location. This has been demonstrated for example through studies by Hershey (1978).

The choice of reference stars should be guided by considerations of location and of magnitude and color compensation between central and reference stars. To reach greatest positional accuracy the configuration should be small in order to minimize film, field and refraction effects. On the other hand for a star of large proper motion, observed over a long interval of time, a larger configuration may well be required. Generally a suitable compromise is aimed for. As mentioned before, on the basis of geometric considerations there is little reason to use more than about six reference stars unless one wants to introduce a non-linear term or terms involving magnitude and color or wishes to check on the astrometric behaviour of the reference stars. Further details may be found in Chapter 7 of *Principles of Astrometry* (van de Kamp, 1967).

The attainable observational accuracy, as well as the problem of systematic errors, has been mentioned in the foregoing Chapters 2 and 3.

CHAPTER 5

PATH OF SINGLE STAR. RELATIVE PARALLAX, PROPER MOTION, QUADRATIC TIME EFFECT

Long-focus photographic technique was first introduced for the determination of annual parallaxes, and continues to be used for that purpose. Schlesinger began this work at the Yerkes Observatory in 1903, and other observatories gradually followed his example: among the first were the Sproul, McCormick, Allegheny, Mount Wilson, Dearborn, and Van Vleck observatories in the United States, the Greenwich Observatory in England, the Cape and the Yale observatories in South Africa, the Stockholm Observatory in Sweden, and the Bosscha Observatory in Indonesia; later additions have been the USNO and the Lick Observatory in the United States. With the exception of Mount Wilson, Bosscha, and Flagstaff, the telescopes are refractors, visual or photographic. Although most parallax determinations so far have been limited to stars brighter than the twelfth magnitude, the programs started with the Lick refractor and especially with the United States Naval Observatory reflector in Flagstaff extend toward fainter stars.

(a) *At the telescope.* Parallax work requires stringent hour-angle restrictions. At the Sproul Observatory, photographs are rarely taken more than half an hour off the meridian. A survey of 30 parallax determinations (van de Kamp and Lippincott, 1950), involving some 3000 plates, reveals little range in the average hour-angle at different times of the night; the algebraic average is within one minute of time from the meridian. The maximum systematic difference in hour-angle, comparing dawn and dusk observations, amounts to less than twenty minutes of time. The corresponding differential dispersion effect in right ascension is slightly more than 0″.001 for an extreme difference in spectrum, A to M, (Chapter 2). This material therefore is systematically well compensated for dispersion effects.

Besides equalizing the magnitude of the central star, whenever required, with the weighted average magnitude of the reference stars, it is desirable to have a small dispersion in the magnitudes of the reference stars. No magnitude error needs generally be feared if the compensation between central star and reference stars is below half a magnitude.

At most observatories plates are taken near extreme parallactic displacement only (shortly after dusk and before dawn) and often only right ascension measurements are used, since they cover the greater part of the parallactic displacement. However, these temporal restrictions may be ignored, thereby permitting parallax determinations, be they less accurate, from declination measures; for other than parallax determinations the restrictions are of no value.

(b) *Plate weight; plate, night, year, measurement errors, double plates, night weights.* Increase in positional accuracy on any one night at the Sproul Observatory is obtained by increasing the number of exposures per plate and the number of plates. Up to four plates are taken on any one night; each plate contains from one to five exposures, spaced one millimeter, or more, apart; plate weights are assigned, depending on quality and number of exposures, according to the following schedule (Table 5.1), based on long experience.

TABLE 5.1
Adopted plate weights

Image quality	No. of Exposures				
	5	4	3	2	1
Good	1.5	1.4	1.2	1.0	0.7
Fair	1.3	1.2	1.0	0.7	0.5
Poor	1.1	1.0	0.7	0.5	0.3

This weighting system takes into account the effect of a *plate error*, common to the successive exposures on any one plate and mostly due to emulsion shifts. Reduction of film errors is obtained by the use of *double* plates, obtained by turning the same photographic plate 180° in its own plane between the two successive sets of exposures, representing the two 'single' plates. The two sets of exposures for the central star are close together on the emulsion; here the effects of a film shift are virtually equal and are opposite for the two successive sets of exposures. The double-plate procedure is especially desirable for large configurations, where plate error is to a great extent responsible for the decreased positional accuracy of single sets of exposures.

The useful number of plates on any one night is limited by the *night errors*, which may be due to refraction anomalies. These errors are assumed constant during all the exposures of the same field within the same night. For the Sproul Observatory Gustav Land (1944) has found an average value of $\pm 0''.0121$ (probable error) for the night error; generally it is not worthwhile to obtain more than two double plates on any one night.

Measurement errors, markedly reduced by improved measuring machines, have been discussed in Chapter 2.

The law of diminishing returns, as enforced by the night errors, makes it advisable to introduce a *night weight*. For example, total plate weights of 5 and 10 may be reduced to effective night weights of only 3 and 4 because of the existence of night errors. A positional value about 0.5μ or $0\overset{''}{.}01$, may be regarded as the ultimate limiting accuracy for any one night. The Sproul weights are discussed by Lippincott (1971, 1974).

Often it will be desirable to obtain a very high accuracy, for example, in studying the systematic behavior of residuals from a solution for parallax and proper motion, due to orbital motion. In this case the positional results obtained on several nights may be combined into average values, referred to as normal points, or normal places. Again, the law of diminishing returns indicates a *'year error'* of $0\overset{''}{.}002$ or more for Sproul observations, a value which appears to set a limit to the ultimate accuracy obtainable in any one year (Chapter 3).

(c) *Analysis for relative parallax, proper motion and quadratic time effect.* As mentioned in Chapter 4, dynamic acceleration effects of the galactic field on an isolated star (multiple star, or cluster) are far too small to be observable at present; such a truly single star has a uniform rectilinear space motion. In practice, over a short time interval, this also may be said of the path of the component of a visual binary with a long period of the order of several decades or centuries: any non-linear orbital effect over a few years would be negligible.

The path of a single star is represented by the following equations of condition:

$$\begin{aligned} X &= c_X + \mu_X t + q_X t^2 + \pi P_\alpha, \\ Y &= c_Y + \mu_Y t + q_Y t^2 + \pi P_\delta, \end{aligned} \tag{5.1}$$

where X and Y represent the reduced position of the central star in right ascension (reduced to a great circle) and declination, measured and reduced to the *standard frame* (Chapter 4) based on the reference stars. The standard frame holds for the equator of a convenient epoch, for example, the year 2000. The first three terms on the right-hand side of the equations represent the heliocentric path defined by the position c_X, c_Y at a zero epoch, the yearly proper motion μ_X, μ_Y and the quadratic time coefficient q_X, q_Y i.e. half the observed acceleration. The latter may be significant for observations on nearby stars covering several decades. The time t is counted from a convenient epoch; the unit of time is the solar or Besselian year, which begins when the right ascension of the mean sun is 280°. The Besselian fraction of the year, τ, is listed, for example, in the American Ephemeris and Nautical Almanac. The fourth term represents the parallactic displacement of the relative heliocentric or annual parallax π in the respective coordinates. The heliocentric parallax factors P_α and P_δ represent the projected fractional equatorial coefficients of the unforshortened angular value of one astronomical unit at the location of the star.

The heliocentric parallax factors in right ascension (reduced to great circle measure) and declination, respectively, are as follows:

$$\begin{aligned} P_\alpha &= R(\cos\varepsilon\cos\alpha\sin\odot - \sin\alpha\cos\odot) \\ P_\delta &= R[(\sin\varepsilon\cos\delta - \cos\varepsilon\sin\alpha\sin\delta)\sin\odot - \cos\alpha\sin\delta\cos\odot] \end{aligned} \tag{5.2}$$

where R is the radius vector of the Earth's orbit expressed in astronomical units, $\varepsilon = 23°27'$ the obliquity of the ecliptic, and $\odot$ the Sun's true longitude; α and δ are the right ascension and declination of the star. The formulae for the parallax factors may be simplified by the following substitutions:

$$\begin{aligned} p &= +0.9174\cos\alpha, \quad a = +0.3979\cos\delta - 0.9174\sin\alpha\sin\delta, \\ q &= -\sin\alpha, \qquad\qquad b = -\cos\alpha\sin\delta \end{aligned}$$

whence

$$\begin{aligned} P_\alpha &= R(p\sin\odot + q\cos\odot), \\ P_\delta &= R(a\sin\odot + b\cos\odot). \end{aligned} \tag{5.3}$$

The positions are oriented close to the equatorial coordinate system of, say, the year 2000. Hence in the calculations of P_α, P_δ the position α, δ is reduced to the equator and equinox of the year 2000; any appreciable effect of proper motion is applied up to the epoch of the observation. A precession correction of $+0\overset{\prime}{.}838$ (2000 − epoch) is applied to the values of $\odot$, to refer them also to the equinox of the observations.

The parallax factors refer to the parallactic displacement with respect to the Sun. The Sun is subject to perturbations from the planets and in principle it is desirable and more accurate to refer the displacement to the barycenter of the solar system. The determination of 'heliocentric' parallaxes is not affected by the perturbations of the sun with respect to the barycenter since the comparatively long periods of the perturbations from the planets in no way interfere with the annual parallactic effect. The matter will be further discussed in Chapter 9.

The values of μ_X, μ_Y, q_X, q_Y and π, obtained from a general least squares solution, are *relative* to the dependence weighted mean background of reference stars. Reduction to absolute is the subject of the next chapter.

(d) *Attainable accuracy.* The ultimate attainable precision of proper motion, quadratic time effect and parallax is the resultant of the errors of the relative values of these quantities and the corresponding reductions to absolute. Let us first consider the attainable accuracy for the relative values of proper motion and quadratic time effect. The same accuracy exists for the rectangular components as for the resultant total proper motion μ and total quadratic time effect q, at least for the case of equal accuracy in the two coordinates, which generally, but not necessarily, would be the case. Assuming *uniform distribution* in time of the observational material over an interval of t years, the weight of the proper motion is approximately proportional to t^3, that of the quadratic time effect to t^5 (Section (e)).

For example, yearly normal points with a probable error 0.″01, over an interval of twenty-five years would yield an error of about 0.″0001 for μ_X, μ_Y, about 0.″00001 for q_X, q_Y; longer intervals would yield corresponding smaller errors. The same yearly accuracy maintained over an interval of fifty years would yield an error of about 0.″00001 for μ_X, μ_Y, well below 0.″000001 for q_X, q_Y. Systematic errors or 'breaks' in technique (instrumental equation, Chapter 3) generally would have little effect on the determination of proper motion and quadratic time effects.

A parallax determination based on some twenty to thirty plates, each with two or three exposures taken at extreme parallactic shifts, extending over a few years, yields a probable error of 0.″01 or less for the relative parallax. By increasing the observational material, and from multiple determinations at different observatories, higher accuracy may be reached with probable errors down to 0.″002 or even less.

Greatest parallactic shifts are obtained shortly after dusk and shortly before dawn. Whereas most parallax information is obtained from measurements in the right-ascension coordinate, it is becoming customary to measure the declination coordinate also. On the average, the weight of parallax determinations in declination is only about fifteen percent of that of determinations from right ascension measures.

Long range systematic errors would hardly influence parallax determinations. The latter might be affected, however, by systematic errors with the not unlikely period of one year, and possibly changing from year to year. It may be considered a source of some reassurance that, on the average, Sproul results yield a negligible difference between determinations of relative parallax from the two coordinates (Chapter 3).

(e) *Calculation of accuracy of quadratic time effect.* To evaluate the attainable accuracy for the quadratic term, we use heliocentric positions. The simplest case is that of three positions, equidistant in time, i.e., three equations over a total interval $2t$ years, and of equal (unit) weight. Choosing the central observation at the time-origin, the unknowns c, μ, q appear as follows in the three equations of condition:

$$c - \mu t + qt^2$$
$$c \qquad (5.4)$$
$$c + \mu t + qt^2$$

The weight of q is $\frac{2}{3}t^4$, corresponding to a probable error of $(\frac{3}{2})^{1/2}t^{-2}$ times the probable error r of one equation. For example, three positions over 50 years, and with $r = 0''.01$ would yield a probable error $0''.00002$ for q or $0''.00004$ for the corresponding acceleration coefficient.

Next we consider the general case of a large number of positions spread uniformly over a total time interval t. Generally the quadratic coefficient is obtained from a least squares solution from numerous individual nights. In order to evaluate the increase in accuracy with time we use a number of normal points equidistant in time, separated by unit time over the total time interval t. We have $(t + 1)$ equations, ranging over the interval 0 to t. The unit of time typically might be one year, but it may be two, or five years, or any other amount. A least squares solution yields the following weight for the quadratic coefficient:

$$[t^4] - \frac{[t^2]^2}{t+1} - \frac{\left([t^3] - \frac{[t]\,[t^2]}{t+1}\right)^2}{[t^2] - \frac{[t]^2}{t+1}}, \qquad (5.5)$$

where the summations extend from 0 to t. Or since

$$[t] = \frac{t(t+1)}{2}$$

$$[t^2] = \frac{t(t+1)(2t+1)}{6}$$

$$[t^3] = \frac{t^2(t+1)^2}{4}$$

$$[t^4] = \frac{t}{30}(t+1)(2t+1)(3t^2+3t-1)$$

expression (5.5) may be reduced to the polynomial form:

$$t(t+1)\left[\frac{[(2t+1)(3t^2+3t-1)}{30} - \frac{t(2t+1)^2}{36} - \frac{t^2(t+2)}{12}\right]. \qquad (5.6)$$

If t is sufficiently large, the above expression, to a high degree of approximation, becomes

$$\frac{t^5}{180}, \qquad (5.7)$$

by neglecting all terms of lower powers. The validity of this approximation depends on the number of conditional equations, *not* on the value of the interval of time. The value from the fifth power approximation is always less than the polynomial; while a very poor approximation for small values of t, the difference drops below 10% for $t = 50$, below 5% for $t = 100$, and to 1% for $t = 500$.

TABLE 5.2
Weight of quadratic coefficient; error of acceleration

t	True weight	Fifth power approximation	Percentage difference	p.e. acceleration from yearly positions with p.e. 0″.01
5	37	17	−54%	0″.003 2
10	858	555	−35	0″.000 68
20	224×10^2	177×10^2	−21	0″.000 13
50	191×10^4	173×10^4	− 9.2	0″.000 014
100	583×10^5	555×10^5	− 4.8	0″.000 002 6
200	182×10^7	177×10^7	− 2.5	0″.000 000 46
500	175×10^9	173×10^9	− 1.0	0″.000 000 048

Table 5.2 gives for the interval t (column 1) the values of the weights of the quadratic coefficient computed from the polynomial (2), from the fifth-power approximation (3), while the fourth column gives the percentage deficiency of (3) below (2). Column (5) gives the probable error of the yearly acceleration (twice the quadratic coefficient), based on a probable error of 0″.01 for successive yearly positions. The idealized situation of equally precise positions would hardly ever be realized, of course. However the fact remains that over a longer time interval with equal observational coverage from year to year, the weight of the acceleration increases approximately with the fifth power of time, a convenient factor to keep in mind. It is obviously only a matter of time to obtain a precise determination of the observed relative acceleration.

CHAPTER 6

REDUCTION TO ABSOLUTE. ACCURACY: COSMIC ERRORS

(a) *Dependence background of reference stars; spurious acceleration.* In order to obtain absolute values for proper motion, acceleration and parallax of the central star, allowance has to be made for the effect of the motions and parallaxes of the reference stars. Their accelerations may be ignored but their proper motions may have an appreciable effect on the acceleration of the central star. If assumed to be known, the proper motions of the reference stars may be eliminated from their measured positions before the path of the central star is analyzed. An alternative, more explicit and elegant method, is to carry out this elimination with the simple aid of dependences, after an analysis of the stellar path; this also permits simple adjustment as improved knowledge of the proper motions of the reference stars may become available.

To reduce μ_X, μ_Y, q_X, q_Y, and π to absolute values, allowance must therefore be made for the *dependence mean background* of the reference stars. A series of astrometric plates taken over a long time interval involves secular terms, owing to the proper motions of both central and reference stars. The proper motion of the central star results in linear time changes in the dependences of the reference stars; these changes in turn result in a linear time change in the dependence-mean proper motion of the system of reference stars. The result is a quadratic time effect, a *spurious acceleration,* which must be carefully evaluated in order to distinguish it from any true secular motion effect or from a long-range perturbation. In long-range problems knowledge of the proper motions of the reference stars is therefore mandatory. In addition there is a secular change in parallax, which generally is of little concern.

(b) *Reduction to fixed background.* The effect of the proper motion and parallax of the reference stars on the reduced position of the central star changes with the time along the stellar path. The effect is most simply expressed through dependences rather than through plate constants. The plate reduction to a standard frame (x_s, y_s) is

$$X = [Dx_s] + X' - [Dx'], \qquad Y = [Dy_s] + Y' - [Dy']. \quad (4.16)$$

Here X', Y' represents the measured position of the central star. The measured positions x', y' of the reference stars may be assumed to have been made in the coordinate system defined by the positions x_0, y_0 of the reference stars at an arbitrary zero epoch conveniently near the temporal center of the series of observations. Hence, in the usual notation, for each reference star

$$x' = x_0 + \mu_x t + \pi_r P_\alpha,$$
$$y' = y_0 + \mu_x t + \pi_r P_\delta, \qquad (6.1)$$

where μ_x, μ_y, and π_r are the proper motion and parallax of the reference star.

For an ideal fixed reference system (x_0, y_0), with neither proper motion nor parallax, the reduced position would be

$$[Dx_s] + X' - [Dx_0], \qquad [Dy_s] + Y' - [Dy_0]. \qquad (6.2)$$

The reduced position (4.16) of the central star requires, therefore, a *correction*

$$+[D(x' - x_0)], \qquad + [D(y' - y_0] \qquad (6.3)$$

to refer the central star to an ideal "fixed" background. This effect may be written as

$$+t[D\mu_x] + P_\alpha[D\pi_r], \qquad +t[D\mu_y] + P_\delta[D\pi_r]. \qquad (6.4)$$

The first term represents a uniformly accelerated motion

$$+t[D_0\mu_x] + t^2[\varDelta D\mu_x], \qquad + t[D_0\mu_y] + t^2[\varDelta D\mu_y], \qquad (6.5)$$

where D_0 are the dependences for $t = 0$ and $\varDelta D$ the annual changes of the dependences. This quantity (6.5) is the motion of the reference background, $+[D_0\mu_x]$, $[D_0\mu_y]$ yearly, due to the dependence mean proper motion at zero epoch, plus the secular acceleration of this background, amounting to $+2[\varDelta D\mu_x]$, $2 [\varDelta D\mu_y]$ yearly.

The second term of expression (6.4) represents a uniformly changing annual parallactic effect

$$P_\alpha[(D_0 + t\varDelta D)\pi_r], \qquad P_\delta[(D_0 + t\varDelta D)\pi_r] \qquad (6.6)$$

due to the dependence mean parallax at the zero epoch, $+[D_0\pi_r]$, plus the yearly secular change of this mean parallax, $+[\varDelta D\pi_r]$.

The *reductions to absolute* are therefore as follows. The relative proper motion μ_X, μ_Y for epoch $t = 0$ may be reduced to absolute by adding the quantity $[D_0\mu_x]$, $[D_0\mu_y]$. The relative quadratic time effect q_X, q_Y is reduced to absolute by adding the quantity $[\varDelta D\mu_x]$, $[\varDelta D\mu_y]$. The resulting absolute quadratic time-effect is half the (absolute) acceleration $d\mu_X/dt$, $d\mu_Y/dt$; the latter is obtained by adding $2[\varDelta D\mu_x]$, $2[\varDelta D\mu_y]$ to the observed, relative, $2q$. The relative parallax π is reduced to absolute by adding $[D_0\pi_r]$; its secular change is generally negligible.

For our present problem the reduction to absolute for the proper motion of the central star generally is a negligible fraction of these quantities, since μ_X, μ_Y are relatively large for most stars of individual astrometric interest. However, the reduction to absolute quadratic time effect may be appreciable, of the same order as the measured quadratic time coefficient q_X, q_Y. Knowledge of the reduction to absolute is therefore vital to obtain the absolute acceleration, and a *sine qua non* is adequate knowledge of the proper motions of the reference stars (Chapter 8).

(c) *Observational and cosmic errors.* How to obtain values of the proper motions and their errors, for the reference stars? At this stage of the art these are best determined by measuring the reference stars on a background of as many stars as possible, in the hope that these represent a fair approximation to a 'fixed' background. The accuracy of proper motions obtained from long-focus photographic plates is limited by two factors:

The *error of observation* r, which depends on (1) the interval between early and recent plates (assuming no intermediate observations), (2) the number of plate pairs, (3) the basic positional accuracy of one plate. This error of observation may be obtained from the inter-agreement of measures for different pairs of plates. Currently it is possible to reduce the probable error of observation r to well below $0\overset{''}{.}001$.

A more serious matter is the *cosmic error* r_{co}, due to the inherently limited number of background stars of sufficiently small motion – as good an approximation as possible of a 'fixed' background – that are suitable for determining the plate constants. At best, a selected group of background stars still has a measurable dispersion in proper motion, relative to their mean. Following Kapteyn (1911), these motions may be referred to as 'cosmic errors'. Their effect on the plate constants is analogous to the inherent uncertainty of the precession constants due to the proper motion components of the observed annual variations in the equatorial positions of the stars used to determine these constants (van de Kamp, 1939).

With present attainable accuracy of observation, the influence of the cosmic errors is not at all negligible. We shall study their effect on the proper motions of stars in a limited photographic field particularly, on the observed acceleration in the proper motion of a central star, referred to the available background, normally via a limited number of reference stars.

We shall limit ourselves to the case of linear plate constants. The measured differences, m_x and m_y, between the photographic positions measured on recent and early epoch plates, reduced to yearly values, are

$$\begin{aligned} m_x &= c_x + a_x x + b_x y + \mu_x + \delta_x, \\ m_y &= c_y + a_y x + b_y y + \mu_y + \delta_y. \end{aligned} \tag{6.7}$$

Here $\mu_x(\mu_\alpha \cos \delta)$, $\mu_y(\mu_\delta)$ are the yearly proper motions in seconds of arc relative to specific plate constant structures c_x, a_x, b_x and c_y, a_y, b_y; δ_x, δ_y are the accidental errors of observation; the (rounded off) rectangular coordinates x and y may be expressed in millimeters. To arrive at values of μ_x and μ_y, we should, ideally, have an extragalactic reference system, or a sufficient number of stars with accurately known proper motions. Since neither of these requirements are likely to be fulfilled at the present time, we use an inter-galactic reference system of n stars selected for small motions relative to their average; these furnish the conditional equations

$$m_x = c_x + a_x x + b_x y,$$
$$m_y = c_y + a_y x + b_y y. \tag{6.8}$$

A least-squares solution yields values for c_x, a_x, b_x and c_y, a_y, b_y.

The rectangular coordinates x and y are referred to the origin of the reference system of the n stars; in this case, highest weight is obtained for c_x and c_y. The values and accuracy of a_x, b_x, a_y, and b_y are independent of the choice of origin.

The probable error of one equation, in each coordinate, is given by

$$R = 0.6745\, \{([\mu^2] + [\delta^2])/(n - 3)\}^{1/2}. \tag{6.9}$$

The probable errors of the plate constants a and b in each coordinate are given by

$$r_a = R/(p_a)^{1/2}, \qquad r_b = R/(p_b)^{1/2}, \tag{6.10}$$

where p_a and p_b are the weights of a and b.

Applying the plate constants to the measured values m_x, m_y, proper motions are obtained relative to the origin of an internal reference system represented by a plate constant structure (c, a, b) in each coordinate. The values of R, and hence the probable errors r_a and r_b of the plate constants, depend on both $[\mu^2]$ and $[\delta^2]$. For a short time interval, the effect of $[\delta^2]$ is likely to be the decisive factor. For large intervals, $[\mu^2]$ determines and limits the attainable accuracy of the plate constants. No matter how small the errors of observation (δ_x, δ_y), it is seen from Equation (6.9) that the accuracy of the proper motions is ultimately limited by the cosmic errors, i.e. the dispersion in the proper motions (μ_x, μ_y) of the background stars.

The origin or zero-point of the proper motions of the background stars remains undefined; the proper motions are 'relative' to this origin. The background system in each coordinate suffers a linear systematic distortion expressed by the term

$$x\varepsilon_a + y\varepsilon_b, \tag{6.11}$$

which has the dimension of proper motion. Here ε_a and ε_b are the (true) errors of the plate constants a and b; they are partly of cosmic, and partly of observational origin. The motions referred to this background are affected by this systematic error, with the opposite sign. Since ε_a and ε_b remain unknown we can only state that the linear distortion (6.11) introduces in each coordinate a probable error

$$\pm(x^2 r_a^2 + y^2 r_b^2)^{1/2} \tag{6.12}$$

for the relative proper motion of a star with coordinates x and y.

The resulting relative proper motions μ_x, μ_y are affected, therefore, by the probable error of observation and by the probable error due to the cosmic plate constant structure (6.11). As time goes by, and high observational accuracy is obtained for proper motions the effect of the cosmic errors may well become more serious than that of the observational errors and the choice of background stars becomes both more critical and more difficult.

(d) *Accuracy of reduction to absolute quadratic time effect.* The path of the central star, which we are interested in, is measured on a limited background of say, three or more reference stars. The conditional equations are of the form

$$\begin{aligned} X &= c_X + \mu_X t + q_X t^2 + \pi P_\alpha \\ Y &= c_Y + \mu_Y t + q_Y t^2 + \pi P_\delta, \end{aligned} \tag{5.1}$$

where the quadratic time coefficients are half the observed acceleration components. A least-squares analysis of the path (5.1) yields probable errors for the unknowns, which are of *observational* origin.

The quadratic time components require a correction for reduction to absolute of

$$+ [\Delta D \mu_x], \qquad + [\Delta D \mu_y], \tag{6.13}$$

The *observational* probable error of the correction for reduction to absolute is given by

$$r_{\text{ob}} = \pm r[(\Delta D)^2]^{1/2}, \tag{6.14}$$

where r is the probable error of observation of the annual proper motions as obtained from the inter-agreement between different pairs of plates. The value r_{ob} depends on the number of reference stars and their areal extent.

In addition there is the effect of *cosmic error.* The effect, in each coordinate, of the plate constant structure on the annual proper motion of each reference star is

$$-(x\varepsilon_a + y\varepsilon_b). \tag{6.15}$$

Hence the effect in each coordinate on the reduction to absolute quadratic time effect

$$-(\Delta D(x\varepsilon_a + y\varepsilon_b)]. \tag{6.16}$$

Since, by definition (within rounding-off errors)

$$[\Delta Dx] = \mu_X, \qquad [\Delta Dy] = \mu_Y, \tag{6.17}$$

where μ_X, μ_Y is the annual proper motion of the central star, Equation (6.16) may be written as

$$-(\mu_X\varepsilon_a + \mu_Y\varepsilon_b). \tag{6.18}$$

The same result is obtained if the central star throughout the plate series had been referred to *all* the background stars. In this case, the cosmic effect on the proper motion of the central star would be

$$-(X\varepsilon_a + Y\varepsilon_b), \tag{6.19}$$

and the corresponding effect on the reduction to absolute quadratic time effect would again be given by Equation (6.18). This result is obvious since the reduction to absolute for three reference stars in fact refers the central star to the entire background of n stars.

All we can state is that the plate constant structure introduces in each coordinate a probable error

$$r_{\text{co}} = \pm(\mu_X^2 r_a^2 + \mu_Y^2 r_b^2)^{1/2} \tag{6.20}$$

for the quadratic time effect of the central star. This probable error is both of observational and cosmic origin; however, as time goes by, the effect of cosmic error becomes preponderant.

The probable error r_{co} of the reduction to absolute is roughly proportional to the dispersion in the proper motions (cosmic errors) of the background stars, and to the proper motion of the central star. Since the secular perspective acceleration is proportional to the proper motion, we conclude that, *ceteris paribus*, the generally most promising cases of a measurable secular perspective acceleration (Chapter 8) are proportionally most affected by the cosmic error of the reduction to absolute.

The error of the reduction to absolute for the quadratic time effect is likely to be rather larger than the error of the relative value q, and therefore may severely limit the attainable accuracy of the absolute value of the quadratic time effect. Illustrations are given in Chapter 8.

(e) *Reduction to absolute parallax.* The reduction from relative to absolute parallax p, is not too serious a problem yet, but it will become so as higher accuracy is obtained in trigonometric parallax determinations. Half a century ago for reference stars of about visual magnitude 10 a standard statistical reduction of 0.″005 was commonly applied, later on reductions ranging from 0.″002 to 0.″007, depending on galactic latitude, were used. Smaller reductions are required for fainter reference stars such as are used for instance in parallax determinations with the USNO reflector.

A sensible approach at present is to evaluate the parallax of each reference star from its apparent magnitude, spectrum (and proper motion), and thus obtain a more precise reduction to absolute than may be found from statistical considerations only. The secular change of the reduction to absolute parallax, for all practical purposes, is negligible.

CHAPTER 7

PARALLAX RESULTS FOR NEAREST STARS. H-R DIAGRAMS

(a) *Review.* The increase in attainable accuracy permits a gradual extension of the distance limit which may satisfactorily be covered by annual parallax determinations. By considerable effort we may obtain reliable parallaxes up to say 50 parsec. Generally a limit of 20 parsec appears indicated for trustworthy results for the nearest stars while 5 parsec has been a conventional limit for our 'immediate' stellar neighborhood.

The nearest stars within 22 parsec have been the subject of continuing study by Gliese (1956, 1969, 1971, 1978). As of December 1978, Gliese lists 1214 stars within 20 parsec, 1578 stars within 22 parsec. Incompleteness sets in at a few parsec as indicated by the following table:

TABLE 7.1
Relative star density of known stars within 22 parsec

Interval in parsec	Relative star density
0–4	1.00
4–5	0.80
5–10	0.54
10–15	0.43
15–20	0.32
20–22	0.29

Maximum frequency appears to be reached at about absolute photographic magnitude +15; it may be concluded that within 10 parsec less than half of all stars have been found thus far. Most of the still undetected stars must be faint objects with proper motions less than 1″ yearly.

(b) *H-R diagrams.* A color-luminosity diagram for stars nearer than 22 parsec, kindly prepared by Gliese (April 1980), is shown in Figure 7.1. It is limited to stars with parallaxes > 0″.045, with mean errors in visual absolute magnitude M_v not exeeding 0.30 mag.

A similar diagram based on parallaxes of faint stars determined with the USNO reflector is shown in Figure 7.2.

(c) *Stars nearer than 5 parsec.* Table 7.2 gives pertinent data on the very

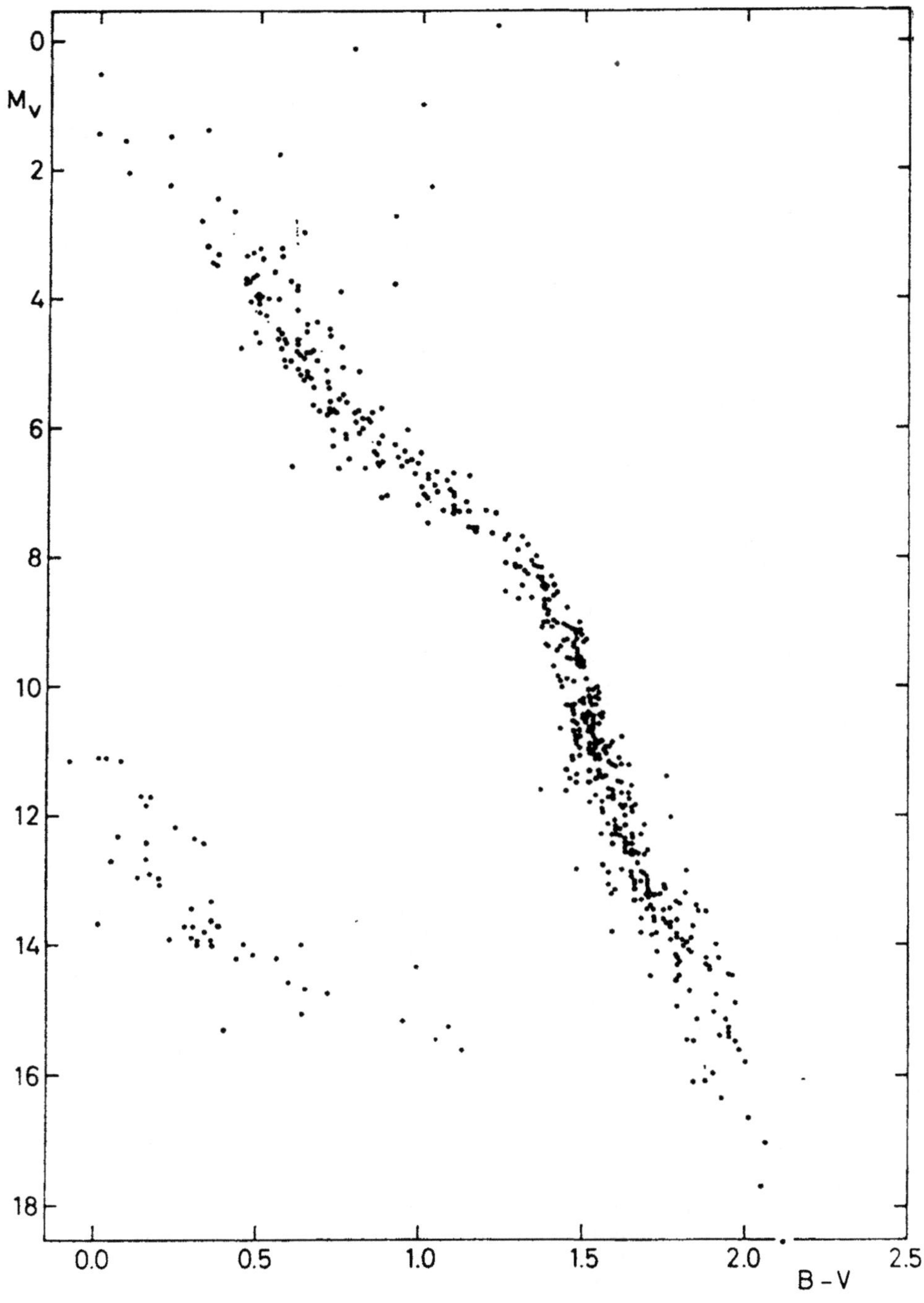

Fig. 7.1. Color ($B - V$)-luminosity (M_v) diagram for stars nearer than 22 parsec (Gliese).

TABLE 7.2

Stars nearer than five parsecs

No.	Gliese No.	Name	(1950)		Proper motion	Position angle	Radial velocity km s^{-1}	Parallax	Distance light years
			RA	Decl.					
1		Sun							
2	559	α Centauri[a]	14^h36^m2	−60°38′	3″68	281°	−22	0″753	4.3
3	699	Barnard's *	17 55.4	+4 33	10.31	356	−108	0.544	6.0
4	406	Wolf 359	10 54.1	+7 19	4.71	235	+13	0.432	7.5
5	411	BD+36°2147	11 00.6	+36 18	4.78	187	−84	0.400	8.2
6	65	Luyten 726-8	1 36.4	−18 13	3.36	80	+30	0.385	8.4
7	244	Sirius	6 42.9	−16 39	1.33	204	−8	0.377	8.6
8	729	Ross 154	18 46.7	−23 53	0.72	103	−4	0.345	9.4
9	905	Ross 248	23 39.4	+43 55	1.58	176	−81	0.319	10.2
10	144	ε Eridani	3 30.6	−9.38	0.98	271	+16	0.305	10.7
11	447	Ross 128	11 45.1	+1 06	1.37	153	−13	0.302	10.8
12	866	Luyten 789-6	22 35.7	−15 36	3.26	46	−60	0.302	10.8
13	820	61 Cygni	21 04.7	+38 30	5.22	52	−64	0.292	11.2
14	845	ε Indi	21 59.6	−57 00	4.69	123	−40	0.291	11.2
15	71	τ Ceti	1 41.7	−16 12	1.92	297	−16	0.289	11.3
16	280	Procyon	7 36.7	+5 21	1.25	214	−3	0.285	11.4
17	725	Σ2398	18 42.2	+59 33	2.28	324	+5	0.284	11.5
18	15	BD+43°44	0 15.5	+43 44	2.89	82	+17	0.282	11.6
19	887	CD−36°15693	23 02.6	−36 09	6.90	79	+10	0.279	11.7
20		G51-15	8 26.9	+26 57	1.26	241		0.273	11.9
21	54.1	L725-32	1 10.1	−17 16	1.22	62		0.264	12.3
22	273	BD+5°1668	7 24.7	+ 5 23	3.73	171	+26	0.264	12.3
23	825	CD−39°14192	21 14.3	−39 04	3.46	251	+21	0.260	12.5
24	191	Kapteyn's *	5 09.7	−45 00	8.89	131	+245	0.256	12.7
25	860	Krüger 60	22 26.3	+57 27	0.86	246	−26	0.254	12.8
26	234	Ross 614	6 26.8	−2 46	0.99	134	+24	0.243	13.4
27	628	BD−12°4523	16 27.5	−12 32	1.18	182	−13	0.238	13.7
28	473	Wolf 424	12 30.9	+9 18	1.75	277	−5	0.234	13.9
29	35	van Maanen's *	0 46.5	+5 09	2.95	155	+54	0.232	14.0
30	1	CD−37°15492	0 02.5	−37 36	6.08	113	+23	0.225	14.5
31	83.1	Luyten 1159-16	1 57.4	+12 51	2.08	149		0.221	14.7
32	380	BD+50°1725	10 08.3	+49 42	1.45	249	−26	0.217	15.0
33	674	CD−46°11540	17 24.9	−46 51	1.13	147		0.216	15.1
34	832	CD−49°13515	21 30.2	−49 13	0.81	185	+8	0.214	15.2
35	682	CD−44°11909	17 33.5	−44 17	1.16	217		0.213	15.3
36	687	BD+68°946	17 36.7	+68 23	1.33	194	−22	0.213	15.3
37		G158-27	0 04.2	−7 48	2.06	204		0.212	15.4
38		G208-44/45	19 53.3	+44 17	0.75	143		0.210	15.5
39	876	BD−15°6290	22 50.6	−14 31	1.16	125	+9	0.209	15.6
40	166	40 Eridani	4 13.0	−7 44	4.08	213	−43	0.207	15.7
41	440	L145-141	11 43.0	−64 33	2.68	97		0.206	15.8
42	388	BD+20°2465	10 16.9	+20 07	0.49	264	+11	0.203	16.0
43	702	70 Ophiuchi	18 02.9	+2 31	1.13	167	−7	0.203	16.0
44	873	BD+43°4305	22 44.7	+44 05	0.83	237	−2	0.200	16.3
45	768	Altair	19 48.3	+8 44	0.66	54	−26	0.198	16.5
46	445	AC+79°3888	11 44.6	+78 58	0.89	57	−119	0.193	16.8
47		G9-38	8 55.4	+19 57	0.89	266		0.190	17.1

[a] The position of α Centuri C ('Proxima') is 14^h26^m3, −62°28′; 2°11′ from the center of mass of α Centauri A and B. The proper motion of C is 3″84 in position angle 282°. Gliese No. 551.

Table 7.2 (continued)

No.	Visual apparent magnitude and spectrum			Visual absolute magnitude			Visual luminosity		
	A	B	C	A	B	C	A	B	C
1	−26.8 G2			4.8			1.0		
2	−0.1 G2	1.5 K0	11.0 M5e	4.4	5.7	15.4	1.5	0.44	0.00006
3	9.5 M5	[b]		13.2			0.00044		
4	13.5 M8e			16.7			0.00002		
5	7.5 M2			10.5			0.0052		
6	12.5 M6e	13.0 M6e		15.4	15.9		0.00006	0.00004	
7	−1.5 A1	8.3 DA		1.4	11.2		23.	0.0028	
8	10.6 M5e			13.3			0.0004		
9	12.3 M6e			14.8			0.00010		
10	3.7 K2	[c]		6.1			0.30		
11	11.1 M5			13.5			0.00033		
12	12.2 M6			14.6			0.00012		
13	5.2 K5e	6.0 K7	[c]	7.5	8.3		0.083	0.040	
14	4.7 K5e			7.0			0.13		
15	3.5 G8			5.9			0.39		
16	0.4 F5	10.7		2.7	13.0		7.0	0.0005	
17	8.9 M4	9.7 M5		11.2	12.0		0.0028	0.0013	
18	8.1 M1eSB	11.0 M6e		10.4	13.3		0.0058	0.00040	
19	7.4 M2e			9.6			0.012		
20	14.8 m			17.0			0.00001		
21	11.5 M5e			13.6			0.00030		
22	9.8 M4			12.0			0.0014		
23	6.7 M0e			8.8			0.025		
24	8.8 M0			10.8			0.0040		
25	9.7 M4	11.2 M6		11.7	13.2		0.0017	0.00044	
26	11.3 M5e	14.8		13.3	16.8		0.0004	0.00002	
27	10.0 M5SB:			11.9			0.0014		
28	13.2 M6e	13.4 M6e		15.0	15.2		0.00008	0.00007	
29	12.4 DG			14.2			0.00017		
30	8.6 M3			10.4			0.00058		
31	12.3 M8			14.0			0.00020		
32	6.6 K7			8.3			0.040		
33	9.4 M4			11.1			0.0030		
34	8.7 M3			10.4			0.0058		
35	11.2 M5			12.8			0.00063		
36	9.1 M3.5SB	[b]		10.8	[b]		0.0040	[b]	
37	13.7 m			15.4			0.00005		
38	13.4 me	14.0 m		15.0	15.6		0.00008	0.00005	
39	10.2 M5			11.8			0.0016		
40	4.4 K0	9.5 DA	11.2 M4e	6.0	11.1	12.8	0.33	0.0030	0.00064
41	11.4 DG			12.6			0.0008		
42	9.4 M4.5			10.9			0.0035		
43	4.2 K1	6.0 K5		5.7	7.5		0.42	0.083	
44	10.2 M4.5e	[b]		11.7	[b]		0.0017	[b]	
45	0.8 A7IV,V			2.3			10.		
46	10.9 M4			12.3			0.0010		
47	14.1 m	14.9 m		15.5	16.3		0.00005	0.000021	

[b] Unseen components.

[c] Suspected.

nearest stars, within 5 parsec; comments on these stars particularly with a view toward unseen astrometric companions (Chapter 13ff) are given elsewhere (Lippincott, 1978).

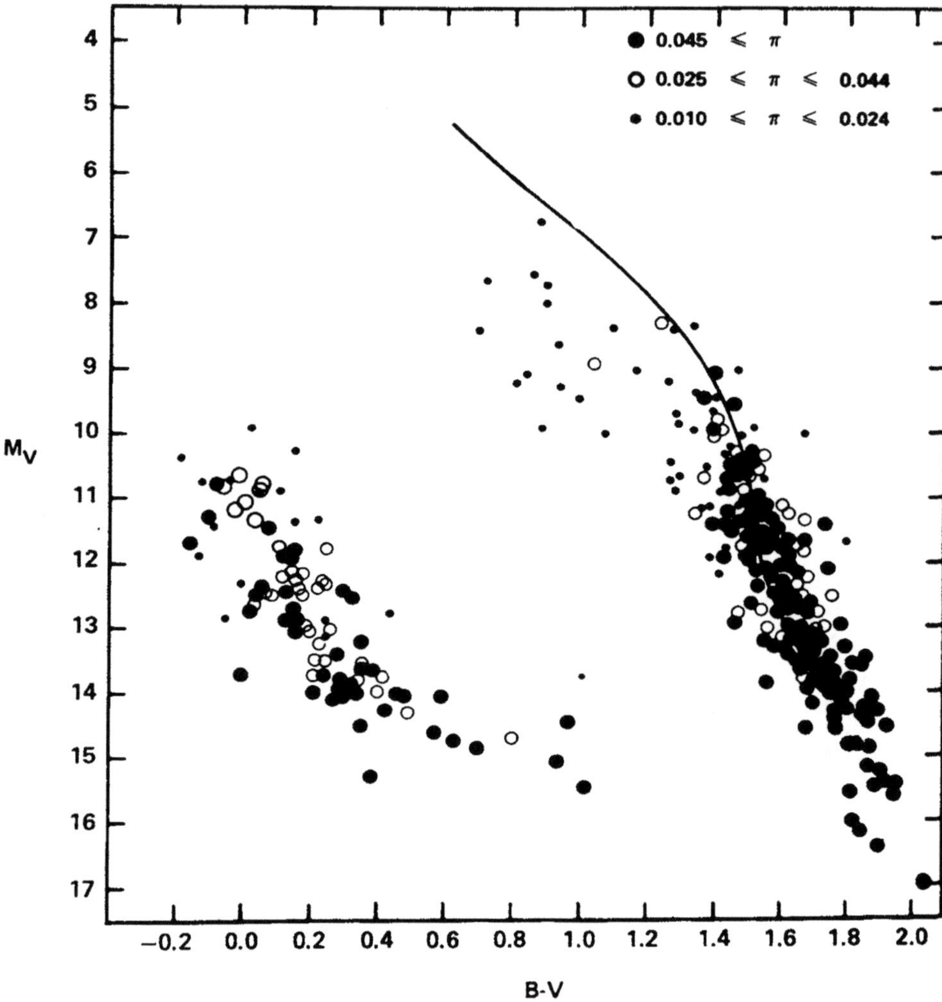

Fig. 7.2. Color-luminosity diagram for faint stars observed with the USNO reflector.

CHAPTER 8

PERSPECTIVE SECULAR CHANGES IN PROPER MOTION, RADIAL VELOCITY, AND PARALLAX

(a) *Introduction.* As time goes by, and stellar paths cover longer time intervals, it becomes necessary in astrometric studies to take into account the perspective secular change in proper motion, which is proportional to the radial velocity V, the (absolute) proper motion μ, and the (absolute) parallax p. Of less immediate concern are the corresponding changes in radial velocity, parallax and apparent magnitude, not to speak of the secular changes in these secular changes.

The secular change in proper motion is of particular interest (a) as a potential method for determining radial velocity, and (b) as a nuisance effect interfering with orbital motion or perturbation. Although we shall primarily by concerned with the secular changes in proper motion we shall record also the formulae holding for the other parameters.

The huge time scale of galactic rotation precludes the detection of deviations from uniform rectilinear motion in the paths of individual stars in our galactic neighborhood for thousands, tens of thousands, or more, years to come. Apart from perturbations due to unseen companions, stellar space motions referred to the Sun, rather to the barycenter of the solar system (Chapter 9) as well as the motions of the barycenters of binary stars for a long time may be considered non-accelerated, i.e., rectilinear at constant velocity.

However the time has come to take into account for the nearest stars secular changes in proper motions, and eventually also in parallaxes and radial velocities, resulting from the changing space-time relation of stars with respect to the Sun. These changes are not of a dynamical nature; they are perspective effects resulting from the change in distance and angle of view.

The secular change in proper motion is proportional to the radial velocity, the proper motion and the parallax. The change in parallax is proportional to the radial velocity and the square of the parallax. The change in radial velocity, always positive, is proportional to the square of the proper motion and inversely proportional to the parallax (section c). The changing values of proper motion, parallax and radial velocity in turn affect the secular changes of these quantities (section d).

Of the three effects, the *secular change in proper motion* is by far the most important, revealing itself through the quadratic time effect within a reasonable time interval because of its rapidly accumulative effect on the stellar path (section e). While the perspective change in proper motion plays a negligible role for observations covering a few years only, after several decades it may become appreciable. If any orbital motion may be ignored or is allowed for, the quadratic time effect eventually may be very accurately determined. This is not to say that the perspective true, absolute, acceleration is necessarily known with the same precision. There is the spurious acceleration in the stellar path caused by the proper motions of the reference stars; the inherent error of this spurious acceleration limits the attainable accuracy of the reduction to absolute (Chapter 6).

Attention was drawn in the beginning of the twentieth century to the possible significance of determining radial velocities independent of Doppler shift, from perspective acceleration in the proper motion of nearby stars of large proper motion (Schlesinger, 1917; Oort, 1932). This eventually may well be the most significant reason for evaluating the acceleration. Illustrations of such a geometric determination of radial velocity are given later (Section f).

(b) *Basic considerations and relations.* Perspective secular changes in astrometric stellar parameters due to space motion are still mostly a matter of academic interest, except for proper motion changes which long ago were anticipated and at least for Barnard's star have now been measured with considerable accuracy (Section e). Formulae for secular changes in stellar proper motion, also radial velocity and parallax have been derived frequently (Bessel, 1844; Seeliger, 1901; Ristenpart, 1902; Schlesinger, 1917; van de Kamp, 1962, 1963b, 1967). The emphasis has always been on the instantaneous rate of change generally involving less than a century, i.e., the first time derivatives. We shall also refer to the behavior of these derivatives over millenia, with particular attention to their maximum absolute values (Section d).

In contrast with the conventional two-body problem we assume no gravitational interaction, i.e., we deal with nonaccelerated motion – uniform rectilinear motion – of the star relative to the Sun, or rather to the barycenter of the solar system (Chapter 9). This is not a dynamical, but a kinematic situation. The polar coordinates of the heliocentric stellar path are the instantaneous distance r at the time τ and the angle or anomaly θ, both counted from perihelion, i.e., closest approach to the Sun. The values of θ range over 180°, from $-90°$ to 0° for approach, 0° to $+90°$ for recession (Figure 8.1).

The present value of the anomaly θ may be derived from the observed values of the radial velocity, V, and the tangential velocity, T, both expressed

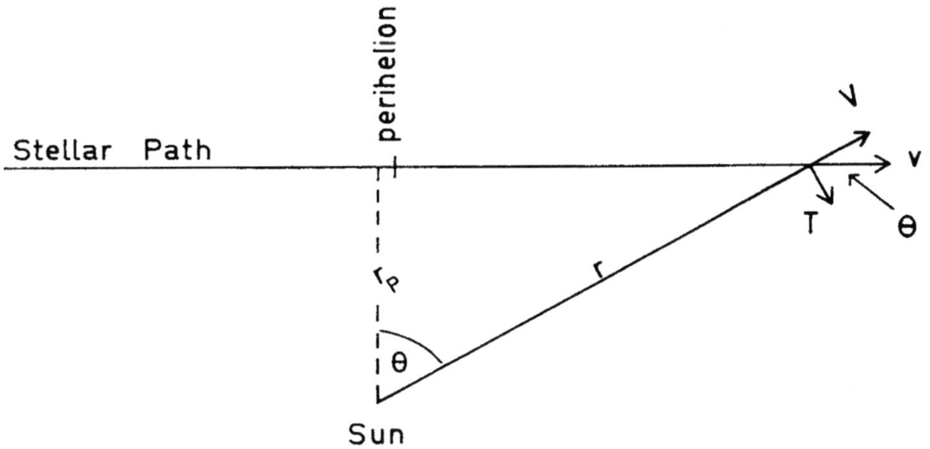

Fig. 8.1. Relation between r, θ, r_P, V, T, and v.

in km s^{-1}. The former is obtained from the observed spectral shift, assumed to be caused by Doppler effect only. The latter is obtained from the observed annual proper motion μ and parallax p through the relation $T = 4.74\mu/p$, since 1 AU yr^{-1} = 4.74 km s^{-1}. The constant space velocity v is given by

$$v = (V^2 + T^2)^{1/2}. \tag{8.1}$$

We have

$$\begin{aligned} V &= v \sin \theta \\ T &= v \cos \theta \end{aligned} \tag{8.2}$$

and

$$\tan \theta = \frac{V}{T}. \tag{8.3}$$

By definition, T is always positive and reaches a maximum at perihelion: V reaches extreme values in the infinite past and infinite future, increases continually, changing from negative to positive at perihelion.

Relations between r, τ, μ, p, and θ. We shall use conventional 'practical' astronomical units, i.e., we express

r in parsecs
τ in years
v, T and V in km s^{-1} as mentioned above
1 parsec = 3.086×10^{13} km
1 yr = 3.156×10^7 s

whence

$$1 \text{ parsec yr}^{-1} = 9.778 \times 10^5 \text{ km s}^{-1}.$$

The kinematical elements of the linear orbit or stellar path in space, are the *constant space velocity* v and the *perihelion distance* r_P. The constant ratio or *path constant*

$$K = \frac{v}{r_P} \tag{8.4}$$

represents the angular velocity at perihelion; it is the equivalent of the constant areal velocity $(r^2/2)(\mathrm{d}\theta/\mathrm{d}t)$ in the dynamical two-body situation.

The perihelion distance is given by

$$r_P = r \cos\theta. \tag{8.5}$$

The distance, in parsecs, travelled since perihelion is

$$r_P \tan\theta. \tag{8.6}$$

Hence the time τ in years from (after $+$, before $-$) perihelion to the present or any chosen epoch is

$$\tau = 9.778 \times 10^5 \, r_P \frac{\tan\theta}{v}$$

or

$$\tau = 9.778 \times 10^5 \frac{\tan\theta}{K} \text{ yr.} \tag{8.7}$$

We may express r as a function of τ; using the identity

$$\cos\theta = [1 + (\tan\theta)^2]^{-1/2}$$

and combining (8.5) and (8.7) we obtain

$$r = r_P \, [1 + (1.023 \times 10^{-6} \, K\tau)^2]^{1/2} \text{ parsec.} \tag{8.8}$$

Over very long time intervals it is convenient to use the argument θ rather than τ; θ has a finite range, while τ runs from $-\infty$ to $+\infty$. A transformation from θ to τ is always possible via (8.7).

The value of the proper motion at any time is given by

$$\mu = \frac{T}{4.74r} = \frac{v\cos\theta}{4.74r}. \tag{8.9}$$

Substituting (8.4) and (8.5) we obtain

$$\mu = \frac{K}{4.74} \cos^2\theta. \tag{8.10}$$

Maximum value $\mu_P = K/4.74$ is reached for $\theta = 0$, i.e., perihelion.

The parallax is given by

$$p = \frac{1}{r} = \frac{\cos\theta}{r_P}, \tag{8.11}$$

maximum and minimum parallax are, obviously, reached for $\theta = 0°$ and $\pm 90°$. More about this in Section (d).

(c) *Changes of μ, V, and p with time.* The perspective changes may be derived in several ways. A simple derivation of the perspective annual change in proper motion has been given by Schlesinger (1917). For the present we have chosen an alternative presentation described elsewhere (van de Kamp, 1962, 1963b, 1967).

For a nonaccelerated motion, the perspective changes perpendicular to and along the radius vector are the Coriolis and centrifugal accelerations in the rotating coordinate system defined by the radius vector joining Sun and star. The secular change perpendicular to the line of sight is

$$-2\frac{dr}{dt}\frac{d\theta}{dt}. \tag{8.12}$$

The angular value as viewed from the Sun is

$$\frac{d^2\theta}{dt^2} = -2\frac{dr}{dt}\frac{d\theta}{dt}\frac{1}{r}. \tag{8.13}$$

The secular change in the line of sight is

$$\frac{d^2r}{dt^2} = +r\left(\frac{d\theta}{dt}\right)^2 \tag{8.14}$$

a quantity which is, of course, always positive.

The secular change in parallax is:

$$\frac{dp}{dt} = -\frac{1}{r^2}\frac{dr}{dt}. \tag{8.15}$$

In these formulae θ is expressed in radians, r in parsec, t in years while $p = 1/r$. To express the yearly secular changes in traditional units of μ, V, and p, recall that the proper motion μ and its annual change $d\mu/dt$ are expressed in seconds of arc, the radial velocity V and its annual change dV/dt in km s^{-1}, and the parallax p and its annual change dp/dt in seconds of arc.

Since 1 parsec = 206265 AU

1 AU yr^{-1} = 4.74 km s^{-1}
1 radian = 206265 arc sec,

we have the following relations:

$$\frac{dr}{dt} = \frac{V}{4.74 \times 206\,265}, \qquad \frac{d^2r}{dt^2} = \frac{dV/dt}{4.74 \times 206\,265}$$
$$\frac{d\theta}{dt} = \frac{\mu}{206\,265}, \qquad \frac{d^2\theta}{dt^2} = \frac{d\mu/dt}{206\,265} \tag{8.16}$$

in traditional units the three relations (8.13), (8.14), and (8.15) become

$$\frac{d\mu}{dt} = -2''.05 \times 10^{-6}\, V\mu p \text{ yr}^{-2}, \tag{8.17}$$

$$\frac{dV}{dt} = +2''.30 \times 10^{-5} \frac{\mu^2}{p} \text{ km s}^{-1}\text{ yr}^{-1}, \tag{8.18}$$

$$\frac{dp}{dt} = -1''.02 \times 10^{-6}\, Vp^2 \text{ yr}^{-1}. \tag{8.19}$$

For completeness we give the annual change in apparent magnitude:

$$\frac{dm}{dt} = +2.17 \times 10^{-6}\, Vp. \tag{8.20}$$

Because of the sign of V, + before, − after perihelion $d\mu/dt$ and dp/dt are positive before, negative after perihelion; dV/dt of course is always positive. Zero values for $d\mu/dt$, dp/dt, dV/dt and dm/dt are reached at infinity, and at perihelion except for dV/dt.

As an illustration: for Barnard's star ($\mu = 10''.31$, $V = -108 \pm 2.5$ km s^{-1}, $p = 0''.547 \pm 0''.003$) we find for the current annual secular changes the following predicted values:

$$\frac{d\mu}{dt} = +0''.001\,25 \pm 0''.000\,03, \qquad \frac{dp}{dt} = +0''.000\,032,$$

$$\frac{dV}{dt} = +0.0046 \text{ km s}^{-1}, \qquad \frac{dm}{dt} = -0.000\,13.$$

(d) *Changes of* $d\mu/dt$, dV/dt, *and* dp/dt *with time or anomaly.* The secular changes of the above secular changes are conveniently formulated through the parameter θ, which has a finite range between −90° and +90°. A transition from θ to τ is always provided by relation (8.7).

Various substitutions lead to

Perspective secular change	Zero value for θ	Absolute maximum value for θ	
$\frac{d\mu}{dt} = -4''33 \times 10^{-7} K^2 \sin\theta \cos^3\theta \text{ yr}^{-2}$	0, $\pm 90°$	$\pm 30°$	(8.21)
$\frac{dV}{dt} = +1.02 \times 10^{-6} vK \cos^3\theta \text{ km s}^{-1} \text{ yr}^{-1}$	$\pm 90°$	0°	(8.22)
$\frac{dp}{dt} = -1.02 \times 10^{-6} \frac{K}{r_P} \sin\theta \cos^2\theta \text{ yr}^{-1}$	0°, $\pm 90°$	± 35.2	(8.23)
$\frac{dm}{dt} = +1.09 \times 10^{-6} K \sin 2\theta$	0°, $\pm 90°$	$\pm 45°$	(8.24)

Maximum (absolute) value for $d\mu/dt$ is reached for

$$\tan\theta = \pm\left(\frac{1}{3}\right)^{1/2} = \pm 0.577 \quad \text{or} \quad \theta = \pm 30° \text{ (exactly)}.$$

This occurs (8.7) for

$$\tau = \frac{\pm 5.6 \times 10^5}{K} \text{ yr}.$$

Maximum value for dV/dt occurs for $\theta = 0°$, i.e. et perihelion. Maximum value for dp/dt is reached for

$$\tan\theta = \pm\left(\frac{1}{2}\right)^{1/2} = \pm 0.707 \quad \text{or} \quad \theta = \pm 35°.2.$$

This occurs (8.7) for

$$\tau = \pm \frac{6.91 \times 10^5}{K} \text{ yr}.$$

The various trigonometric functions are plotted in Figure 8.2. For completeness we note that the secular change in magnitude is zero at infinity and at perihelion. Maximum (absolute) change for dm/dt is reached for $\theta = \pm 45°$; this occurs for

$$\tau = \frac{\pm 9.778 \times 10^5}{K} \text{ yr}.$$

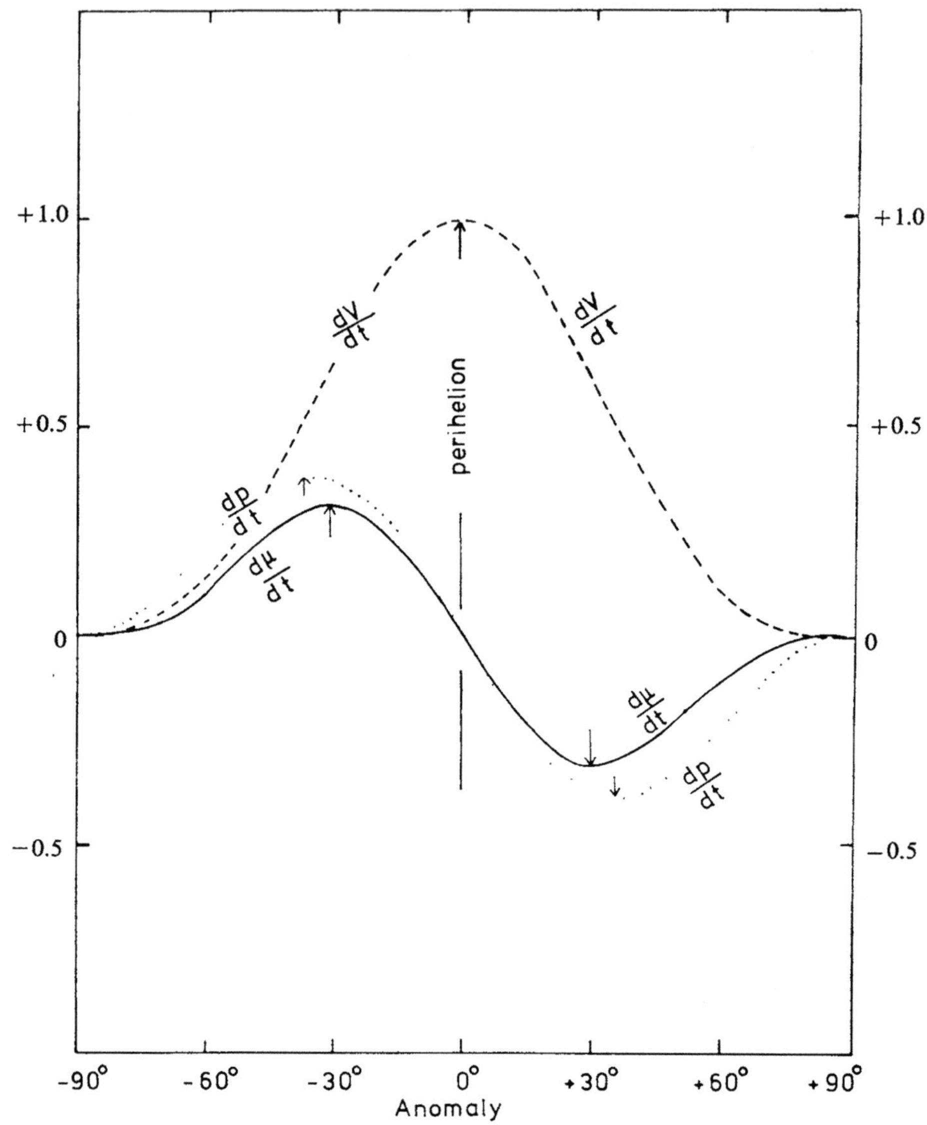

Fig. 8.2. Proportionality factor as function of anomaly for secular changes in $d\mu/dt$, dV/dt, and dp/dt.

(e) *Determination of perspective secular acceleration. Examples: Barnard's star and van Maanen's star.* The most significant of the secular changes, $d\mu/dt$, may be obtained from observation of long-range astrometric series of plates which provide the equatorial components q_X and q_Y of the quadratic time effect, i.e. half the observed acceleration. The equatorial components of the true acceleration $d\mu/dt$ are obtained by adding the reduction to absolute due to the proper motions of the reference stars (Chapter 6), i.e.

$$\frac{d\mu_X}{dt} = 2q_X + 2[\Delta D\mu_x],$$
$$\frac{d\mu_Y}{dt} = 2q_Y + 2[\Delta D\mu_y]. \tag{8.25}$$

The errors of the *observed* acceleration $2q_X$, $2q_Y$ depend primarily on the temporal distribution of the observational material and generally to a much lesser extent on any orbital motion or perturbation. These errors, after several decades, may be reduced to minute values, 0$''$00001 (p.e.) or less; however the reduction to absolute may not be that accurate. Its error depends on the cosmic errors, i.e. the dispersion in the proper motions of available background stars, and *ceteris paribus* is approximately proportional to the proper motion of the central star (Chapter 6). Two examples follow.

(1) *Barnard's star*. 17^h55^m4, + 4°33′, 9.5, M5, p = 0$''$547. Lundmark and Luyten (1922) made a first attempt to determine the acceleration from observations up to 1919, including an early observation made by Lamont in 1842 with the Munich meridian circle (Crommelin, 1916). Alden (1924)

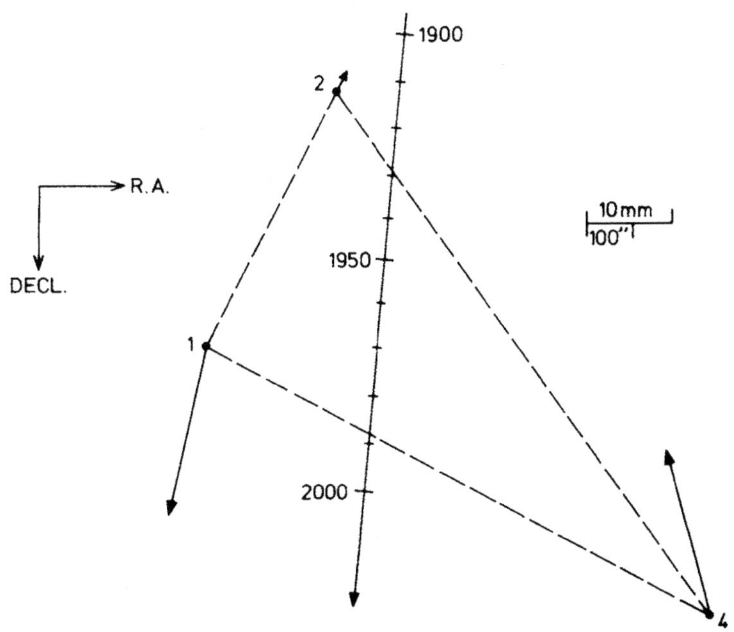

Fig. 8.3. Path of Barnard's star. Reference stars and their proper motions in 10 000 yr.

made a photographic study of the proper motion from McCormick photographs covering the interval 1916–24. A subsequent, still provisional attempt was made by the present author (van de Kamp, 1934), but no valid and accurate determinations of the acceleration were made until recently (van de Kamp, 1969a, 1975, 1977a, b), when sufficient time had elapsed. The latest determination now follows. Measurements of Sproul plates taken on 1165 nights over the interval 1916–1919; 1938–79, referred to three reference stars, whose proper motions recently were measured on a background of twenty two stars, yield

		RA (x)	Decl. (y)		
Observed acceleration	$2q$	$+0''.000\,11$	$+0''.000\,60$	000 01	(p.e.)
Reduction to absolute	$2[\Delta D\mu]$	$-$ 22	$+$ 70	3	,,
Absolute acceleration	$\frac{d\mu}{dt}$	$-$ 11	$+$ 130	3	,,
Predicted acceleration		$-$ 10	$+$ 124	3	

The total absolute acceleration is $+0''.001\,30 \pm 0''.000\,03$.

The agreement between observed and predicted is satisfactory, particularly in view of the comparatively large errors for the reduction to absolute.

Determinations from other observatories (McCormick, Allegheny, Van Vleck) range from $-0''.000\,07$ to $-0''.000\,20$ in RA and from $+0''.001\,21$ to $0''.001\,37$ in Decl. (van de Kamp, 1977c).

(2) *Van Maanen's star.* 0^h46^m5, $+5°9'$, 12.4, DG, $p = 0''.232$. For comparison we record also the Sproul results for the secular acceleration of van Maanen's star (Hershey, 1978). Recall that the error of the observed acceleration depends on the density of observational material, which is appreciably higher for Barnard's star than for van Maanen's star and is the principal cause of the much smaller error for the observed acceleration of Barnard's star. On the other hand the error of the reduction to absolute is roughly proportional to the proper motion of the central star and to the dispersion in the proper motions (cosmic errors) of the background stars (Chapter 6). For Barnard's star this error is comparatively large due to the limited number (14) of background stars and the large proper motion ($10''.3$) of Barnard's star.

For van Maanen's star this error is virtually negligible, partly because of the smaller proper motion ($3''.0$) of van Maanen's star but primarily due to the background stars. Because of their larger number (45) and smaller internal cosmic errors they form a much more nearly ideal fixed background, than is the case for Barnard's star.

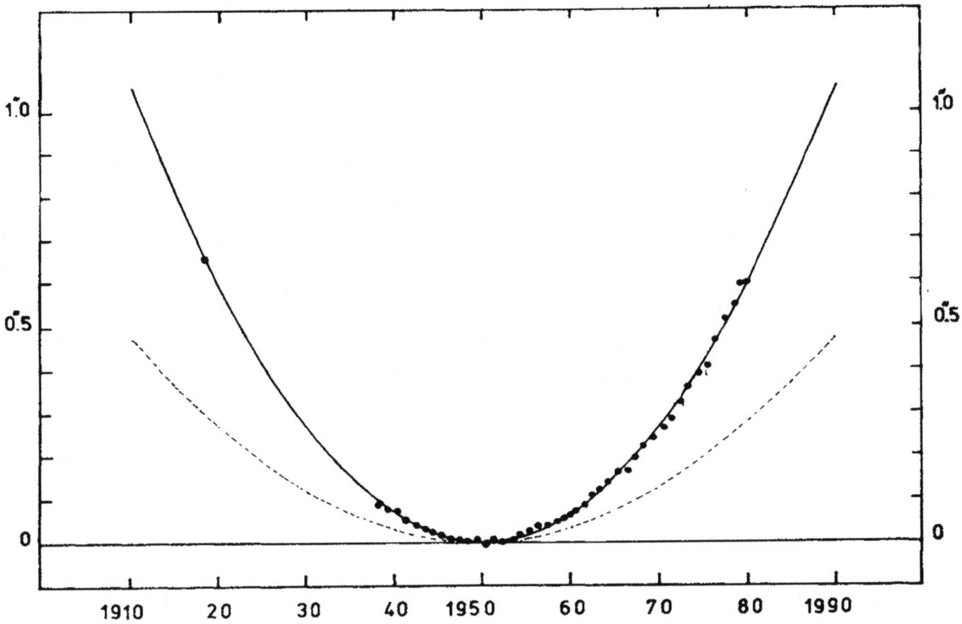

Fig. 8.4. Barnard's star. Secular acceleration in declination. The broken curve illustrates the accumulated observed acceleration effect relative to the epoch 1950.000, as determined from Sproul positions over the interval 1916–1919, 1938–1979. The proper motions of the three reference stars introduce a spurious acceleration, which, with opposite sign, gives the reduction to absolute. The resulting absolute accumulated acceleration effect is given by the solid curve together with normal points (yearly beginning 1938) from the least squares solution (Chapter 16) for parallax, proper motion and quadratic time effect (but not perturbation).

The figure illustrates dramatically the increasing accumulated absolute acceleration effect, which is overwhelmingly large, over 0.″6, compared with the small perturbation (Chapter 16), whose total amplitude is 0.″03.

The errors for the observed acceleration, reduction to absolute, and absolute acceleration (8.25) are as follows:

p.e.	Barnard's Star	van Maanen's Star
observed acceleration $2q$	±0.″000 008	±0.″000 026
reduction to absolute $2[\Delta D\mu]$	±0.″000 027	±0.″000 004
absolute acceleration $\frac{d\mu}{dt}$	±0.″000 028	±0.″000 026

The observed acceleration is vastly more accurate for Barnard's star then for van Maanen's star, while the opposite holds for the reduction to absolute. The combination of the two sources of errors leads to comparable accuracy for the absolute accelerations of the two stars.

(f) *Astrometric determination of radial velocity.* Ultimately one of the significant aspects of perspective secular acceleration is the possibility of determining a value of the radial velocity of a star without recourse to spectral shift; the latter may be affected by other than true Doppler shift, such as gravitational redshift.

Adopting values for μ and p which usually are known with considerable precision, a determination of $d\mu/dt$ yields a value for radial velocity

$$V = -4.88 \times 10^5 \frac{d\mu}{dt} (\mu p)^{-1} \text{ km s}^{-1}, \tag{8.26}$$

the formula to be used for either X or Y coordinate or the resultant of the two.

Two illustrations will be given. The earlier mentioned results for Barnard's star yield $V = -112 \pm 3$ km s^{-1}, in satisfactory agreement with the value -108 ± 2.5 km s^{-1} determined spectroscopically. Improvement of the radial velocity of Barnard's star is to be obtained from either or both astrometric and spectroscopic observations.

The other example is van Maanen's star.

An early determination of radial velocity, $+238$ km s^{-1} (Adams and Joy, 1926), led to the notion that this high value might be caused by a gravitational redshift as large as $+700$ km s^{-1} (Russell and Atkinson, 1931; Shapiro and Teukolsky, 1976). Oort (1932) questioned the reality of a large radial velocity which was incompatible with the theory of galactic rotation; he proposed that the matter could be settled by an astrometric determination of the radial velocity.

Both observed radial velocity and estimated red shift have been drastically revised since. The observed spectral shift, interpreted as Doppler effect, may now be taken as $+39 \pm 4$ km s^{-1} (Greenstein, 1972). Meanwhile astrometric material has accumulated sufficiently to permit a provisional determination of radial velocity from the observed quadratic time effect reduced to absolute, i.e. the secular perspective acceleration.

A first determination from Sproul plates (van de Kamp, 1971) has been superseded. A study by Hershey (1978) from Sproul plates taken on 208 nights over the interval 1939–1976 yields a radial velocity of $+25 \pm 18$ km s^{-1}, leaving only $+14 \pm 18$ km s^{-1} for gravitational redshift, disregarding uncertainties of other physical origins. This result is consistent with the earlier determination of $+33 \pm 11$ km s^{-1} based on Allegheny plates taken over the interval 1917–1972 (Gatewood and Russell, 1974).

Although these results for van Maanen's star have limited accuracy they put a limit on both radial velocity and gravitational redshift which certainly could not be as large as was suspected half a century ago.

(g) *Evaluation and elimination of quadratic time effect.* This becomes increasingly important and necessary for observations made over long time intervals, where the orbital motion has to be separated from the quadratic time effect of perspective and spurious origin. An attempt must then be made to predict $d\mu_X/dt$, $d\mu_Y/dt$ (absolute) from known μ_X, μ_Y, V, and p and to evaluate the reduction to absolute from knowledge of the μ_x, μ_y of the reference stars. This is of particular singificance for a visual binary, in which the orbital motion has to be separated from any non-orbital quadratic time effect.

To eliminate the quadratic time effect, the following values of the predicted quadratic time effect, with opposite sign, have to be added to the observed positions X and Y in each coordinate:

$$\begin{aligned} -q_X &= -\frac{1}{2}\frac{d\mu_X}{dt} + [\Delta D\mu_x] \\ &= +1''.02 \times 10^{-6}\, V\mu_X p + [\Delta D\mu_x], \\ -q_Y &= -\frac{1}{2}\frac{d\mu_Y}{dt} + [\Delta D\,\mu_y] \\ &= +1''.02 \times 10^{-6}\, V\mu_Y p + [\Delta D\mu_y], \end{aligned} \tag{8.27}$$

An example is given later for the analysis of orbital motion and mass-ratio of the visual binary 61 Cygni (Chapter 12).

CHAPTER 9

REDUCTION FROM HELIOCENTRIC TO BARYCENTRIC

(a) *Perturbation of solar path.* The space motion of the sun and hence its reflected secular parallactic component of the star's proper motion are uniform and rectilinear apart from any secular acceleration (Chapter 8); any long term deviation (curvature) resulting from galactic rotation is negligible for millenia to come. However the space motion of the Sun is subject to a multiple perturbation by its planets. Each component perturbation has a period equalling that of the planet, and a total amplitude proportional to the mass of the planet and to its mean distance from the Sun. The principal contributors are the four giant planets, in the first place Jupiter and Saturn, and over longer time intervals Neptune and Uranus. The effects of Pluto and of the terrestrial planets are negligible.

Relevant data for the five outer planets are (Clemence, 1953):

Planet	Mean distance from Sun in AU	Period in years	Inverse mass $\mathscr{M}_{\odot}^{-1}$	Semi-axis major of perturbation in AU
Jupiter	5.20	11.86	1 047	0.0050
Saturn	9.54	29.46	3 512	27
Uranus	19.19	84.01	22 869	8
Neptune	30.07	164.78	19 314	16
Pluto	39.52	248.42	360 000	1

The Sun therefore describes a comparatively minute perturbation orbit around the barycenter of itself and the planets, with a total possible effect of 0.010 AU or 0.010 times the parallax as revealed in angular measure for a star.

What effect does this have on the analysis on a stellar path? A conventional parallax determination involves parallax, proper motion, quadratic time effect and possibly orbital motion. If we use heliocentric parallax factors (Chapter 5) the value of the parallax will not be measurably affected by the Sun's perturbation whose principal component periodicities range from 12 to 165 years. However the heliocentric stellar path differs ever so slightly from the uniform rectilinear path referred to the barycenter of the solar system.

This is a matter of importance in the study of long term orbital phenomena covering several decades, particularly for small perturbations in the path of a nearby star. For example the extreme difference between barycentric and heliocentric parallactic affects is 0".0076 for Alpha Centauri ($p = 0''.753$), 0".0055 for Barnard's star ($p = 0''.547$). Half of this extreme effect would be due to Jupiter alone. In principle therefore the possibility exists, marginally to be true, from intensive and extensive observations to discover at least Jupiter, possibly Saturn, from their perturbations on the Sun's path as reflected in the secular parallactic components of the proper motion of Alpha Centauri and of Barnard's star! In practice we correct for these effects whenever necessary.

It is interesting to reflect that, *ceteris paribus*, were our sun a red dwarf with mass $0.14\mathcal{M}_\odot$, these perturbations would be seven times as large as they actually are. The extreme amplitude would be 0.07 AU, i.e. 0".053 for Alpha Centauri, 0".038 for Barnard's star. (This is a rather academic situation, since most likely we would not be available as potential observers.)

(b) *Heliocentric and barycentric parallax factors.* In traditional annual parallax determinations the parallactic displacements are referred to the Sun as origin and the heliocentric parallax factors calculated accordingly (Chapter 5):

$$P_\alpha = R(\cos\varepsilon\cos\alpha\sin\odot - \sin\alpha\cos\odot),$$
$$P_\delta = R[(\sin\varepsilon\cos\delta - \cos\varepsilon\sin\alpha\sin\delta)\sin\odot - \cos\alpha\sin\delta\cos\odot].$$

Alternate formula for the heliocentric parallax factors are

$$\begin{aligned} P_\alpha &= -X\sin\alpha + Y\cos\alpha \\ P_\delta &= -X\cos\alpha\sin\delta - Y\sin\alpha\sin\delta + Z\cos\delta, \end{aligned} \tag{9.1}$$

where X, Y, Z, are the heliocentric equatorial rectangular coordinates, in astronomical units, of the Sun relative to the Earth:

$$\begin{aligned} X &= R\cos\odot, \\ Y &= R\sin\odot\cos\varepsilon, \\ Z &= R\sin\odot\sin\varepsilon. \end{aligned} \tag{9.2}$$

These coordinates are counted positive toward the vernal equinox, to the direction toward RA = 90° on the celestial equator and to the celestial north pole respectively.

To reduce to barycentric, we use the heliocentric equatorial rectangular coordinates dX, dY, dZ (in astronomical units) of the barycenter of the Sun and planets, relative to the Sun. Their values involving the five outer planets have been calculated for the interval 1800 to 2060 and are published in Astronomical Papers of the American Ephemeris and Nautical Almanac (Clemence, 1953).

Hence with increased observational precision a shift from Sun to barycenter of the solar system should be considered. The following corrections are required to reduce the heliocentric to barycentric parallax factors:

$$\begin{aligned} \Delta P_\alpha &= -\mathrm{d}X \sin\alpha + \mathrm{d}Y \cos\alpha \\ \Delta P_\delta &= -\mathrm{d}X \cos\alpha \sin\delta - \mathrm{d}Y \sin\alpha \sin\delta + \mathrm{d}Z \cos\delta \end{aligned} \tag{9.3}$$

resulting in the barycentric parallax factors:

$$\begin{aligned} {}_bP_\alpha &= -(X + \mathrm{d}X)\sin\alpha + (Y + \mathrm{d}Y)\cos\alpha, \\ {}_bP_\delta &= -(X + \mathrm{d}X)\cos\alpha \sin\delta - (Y + \mathrm{d}Y)\sin\alpha \sin\delta + \\ &\quad + (Z + \mathrm{d}Z)\cos\delta. \end{aligned} \tag{9.4}$$

The barycentric parallax factors are thus obtained by adding $\mathrm{d}X$, $\mathrm{d}Y$, $\mathrm{d}Z$ to X, Y, Z. And for a high precision analysis of stellar orbital periodicities of the order of a year it appears advisable to use barycentric parallax factors *ab initio*. It remains of course possible because of the long range character of the Sun's perturbation to allow for the differential effect from year to year, when one is concerned with long range orbital effects or perturbations. The principal component of this perturbation due to Jupiter has a period of nearly 12 years, the longest annual observational season is not more than half a year. With existing accuracy we may safely adopt one and the same correction for the effect of Jupiter, and of the other planets, for any one observational season, instead of for each night of observation. This is a reasonable and acceptable procedure when a perturbation in a stellar path has been analyzed from annual mean residuals from a first solution for proper motion, heliocentric parallax and quadratic time effect.

(c) *Illustration: Barnard's star.* For Barnard's star ($17^{\mathrm{h}}55^{\mathrm{m}}.4$, $+4°33'$, 1950, $\pi = 0''.55$) the corrections from heliocentric to barycentric parallax-factors (9.3) amount to

$$\begin{aligned} \Delta P_\alpha &= +\mathrm{d}X - 0.02\,\mathrm{d}Y, \\ \Delta P_\delta &= +0.08\ Y + \mathrm{d}Z. \end{aligned}$$

The coefficients ΔP_α, ΔP_δ are expressed in astronomical units of distance. The resulting corrections for heliocentric to barycentric positions are $-\pi\,\Delta P_\alpha$, $-\pi\,\Delta P_\delta$; extreme values are

$$\text{in RA} \begin{cases} -0.0059\ \text{AU} = -0''.0032 \text{ in } 1939, \\ +0.0081\ \text{AU} = +0''.0048 \text{ in } 1957, \end{cases}$$

$$\text{in Decl.} \begin{cases} -0.0038\ \text{AU} = -0''.0021 \text{ in } 1942, \\ +0.0037\ \text{AU} = +0''.0020 \text{ in } 1961. \end{cases}$$

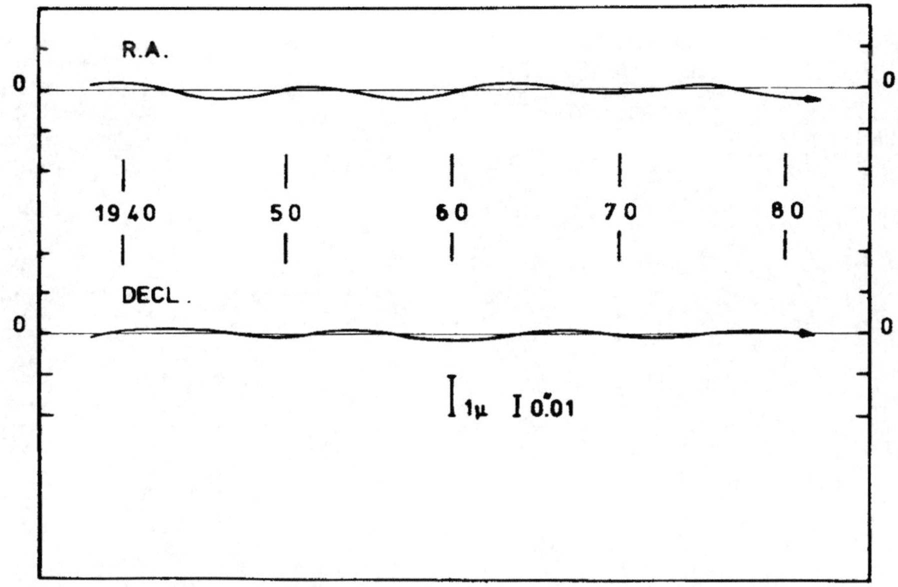

Fig. 9.1. Effect of Sun's perturbation caused by the five outer planets on the position of Barnard's star (17^h55^m4, $+4°33'$, 1950; parallax 0".547). If ignored the results for the perturbation in the path of Barnard's star would be slightly affected.

The total range 0".008 in RA and 0".004 in Decl. is sufficiently large to be taken into account in an analysis of the observed small perturbation with a total amplitude range of 0".021 in RA and 0".026 in Decl. over the interval 1940–1980 (Chapter 16).

Figure 9.1 shows the Sun's perturbation as revealed at the location of Barnard's star over the interval 1938–1981.

The USNO reflector in Flagstaff, Arizona with Dr K. Aa. Strand. Aperture 155 cm, focal length 15.22 m. Scale in focal plane 1 mm = 13.″55 or 1″ = 74 microns.

CHAPTER 10

VISUAL BINARIES; ORBITAL ELEMENTS

(a) *Introduction.* Visual binaries have been discussed by several authors, of whom we mention:

R.G. Aitken: 1935, *The Binary Stars* (Dover Publications).
P. Muller: 1962, 'Techniques for Visual Measurements', *Stars and Stellar Systems*, Vol. II, p. 440.
W.H. van den Bos: 1962, 'Orbit Determinations of Visual Binaries', *Stars and Stellar Systems*, Vol. II, 537.
P. Couteau: 1978, *L'Observation des étoiles doubles visuelles* (Flammarion, Paris).

The present author's earlier astrometric treatise (1967) contains a chapter on visual binaries, which is repeated here with minor changes. We shall be primarily interested in the orbital effects of visual binaries on the paths of the components as measured on a background of reference stars (Chapters 11ff).

Historically the first astrometric discoveries and studies of binaries were made by visual techniques, hence the designation 'visual binaries'. Long-focus telescopes and accessories provide the relative location of the components as seen projected on the sky, i.e. perpendicular to the line of sight. We shall (later) distinguish between resolved and unresolved astrometric binaries depending on whether the components are observed seperated or blended (Chapter 11).

Visual observations of binaries permit the resolution of pairs with a separation well below 1″; the photographic method generally does not yield clearly separated images below about 0.1mm or 2″. It is true that under excellent seeing conditions and for nearly equal magnitudes of the components, apparently well-separated images may be obtained down to 1″. On the other hand, for large magnitude differences, clear separation may not be obtained below about 0.2 mm or 3 or 4″. In all these cases of close proximity of images, or partial overlap, one must reckon with the possibility of appreciable systematic errors in the measurement of the separation, and the results are to be regarded with caution. Even for apparently well-separated images it is not at all certain that systematic 'proximity' effects may not exist up to about 0.3 mm or separations as far as 5 or 6″, as indicated at the Sproul Observatory for the 'wide' binary Xi Bootis (Hershey, 1977).

Hence there is every reason to observe the 'close' double stars, visually, by micrometer or by interferometer, and to limit the photographic observations to wider pairs for which the images are clearly separated on the photographic plate, or use special methods such as 'speckle' photography (McAlister, 1977).

The relative position of the fainter secondary star (B) referred to the brighter primary (A) is yielded by visual observations in polar co-ordinates: distance ρ and position angle θ, by photographic observations in the form of equatorial rectangular co-ordinates ΔX and ΔY.

The following relations exist:

$$\begin{aligned} \Delta\alpha \cos\delta &= \Delta X = \rho \sin\theta, \quad \Delta\delta = \Delta Y = \rho\cos\theta, \\ \rho &= \sqrt{(\Delta X)^2 + (\Delta Y)^2}, \quad \theta = \arctan\frac{\Delta X}{\Delta Y}. \end{aligned} \tag{10.1}$$

To permit a comparison of observations at different epochs all position angles are reduced to a standard equator, say of the year 2000. The position angle at epoch t is corrected to the equator of epoch 2000 by adding $+\,0\overset{\circ}{.}0056 \sin\alpha \sec\delta\,(2000 - t)$ for precession, and $+0\overset{\circ}{.}00417\,\mu_\alpha \sin\delta\,(2000 - t)$ for proper motion, where μ_α is the annual proper motion in right ascension expressed in seconds of time.

(b) *Multiple exposure technique.* High accuracy has been reached with the

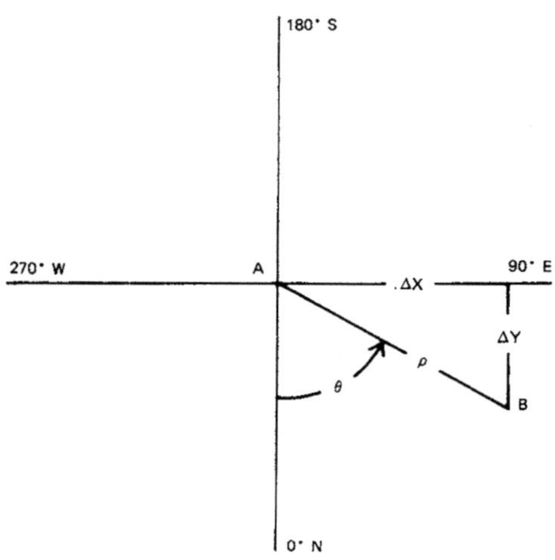

Fig. 10.1. Relative position of double-star components.

photographic multiple exposure technique, succesfully used with long-focus instruments. Precision technique of long-focus photographic measurements of double stars was first developed by E. Hertzsprung (1920), with the visual refractor of 50 cm aperture and 12.5 m focal length at the Potsdam Observatory during the years 1914–1919. This type of observation has been continued by others, notably by K. Aa. Strand (1937, 1946).

In observing the relative positions of the components of a resolved astrometric binary by this method, magnitude error is compensated by the use of a coarse grating in front of the objective. Such a grating produces diffraction images symmetrically located with respect to the central image; these images can be given any desired intensity with respect to the central image by proper choice of the thickness of the bars and of their spacing. Multiple exposures are taken; each plate has two rows of exposures, in a west-to-east sequence parallel to the daily motion. After a series of about thirty to forty exposures has thus been obtained, either manually or by an automatic device, the telescope is given a small shift in declination and the double star itself or a neighboring bright star is used to impress a trail giving the equator of the date. As a rule, a second row of exposures is then taken, followed by a second trail. The exposure times never exceed 30 s; on the other hand, no exposures below 3 s are used. The relative position of the two components is obtained from the difference in $\Delta\alpha \cos\delta$ (ΔX) and $\Delta\delta$ (ΔY), measured, for example, on a precision long-screw machine. For an average plate with forty to fifty measured expo-

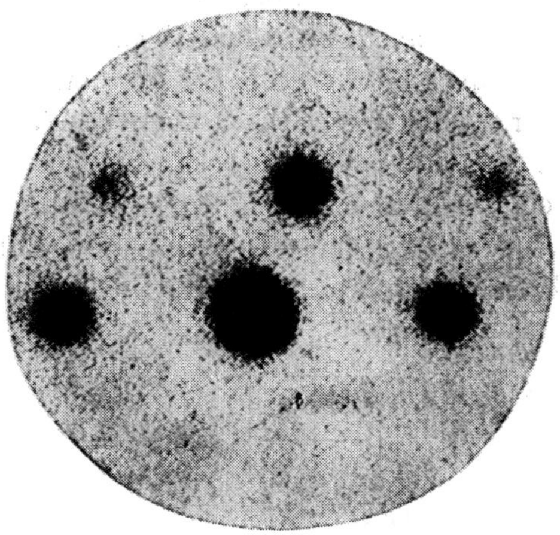

Fig. 10.2. A 5-second exposure of Castor, enlarged 75 ×. The separation of the components is 3.″74, or 0.198 mm on the plate. The first-order spectra are 1 mag. fainter than the central image. Photographed December 1, 1939, by K. Aa. Strand, with the Sproul 61 cm refractor, aperture reduced to 32 cm, Eastman IV-G emulsion, Wratten No. 12 (minus blue) filter. Scale of original photo: 1 mm = 18.″87.

sures, the relative position ΔX, ΔY of the two components is obtained with a probable error of $\pm 0''.006$. Part of this error, about $\pm 0.2\mu$ or $\pm 0''.004$, is personal or instrumental error of measurement; it affects the distances rather than the position angles.

The error in orientation arising from inaccuracies in the trail is below $0^\circ.01$ for any one plate and would therefore be less than $0''.002$ for a pair with a separation of $10''$. A normal point based on 6 or more multiple exposure plates yields a precision of about $0''.002$, i.e. the same as the attainable yearly accuracy in the differential method (Chapter 3).

(c) *Kepler's problem. Elliptical rectangular coordinates.* The location in an elliptical orbit is determined by the three *dynamical elements:* period P, eccentricity e, epoch of periastron passage T, plus the epoch of observation t. This is called Kepler's problem and the solution is easily followed by introducing Kepler's *auxiliary circle*, which is located in the orbital plane tangent to the orbit in periastron and apastron and from which the elliptical orbit may be derived by the foreshortening factor $\sqrt{1 - e^2}$ perpendicular to the major axis. The radius of Kepler's circle is therefore equal to the semi-axis major a of the orbit, and appears on the sky as the semi-axis major of the auxiliary *Kepler ellipse.*

The angular coordinate of the position in the ellipse is the *true anomaly* v counted from periastron in the direction of orbital motion. The position of the corresponding location in the auxiliary circle is measured at the center by the *eccentric anomaly.* E, also counted from periastron in the direction of orbital motion. It is convenient to introduce the *unit orbit* and the *unit circle*, for which $a = 1$. The radius vector FS is the projection of the radius vector FS', which sweeps over equal areas of the circle in equal intervals of time. The area of the unit circle is π, and is swept over in one period P. Hence the area $FS'P$ in the unit circle, covered in the time $t - T$ since periastron passage, amounts to $\pi(t - T)/P = (n/2)(t - T)$ (see p. 69). This area may be expressed as the difference between the sector $S'OP$ and the triangle $OS'F$, i.e.,

$$S'OP - OS'F = FS'P,$$

or

$$\frac{E}{2} - \frac{e \sin E}{2} = \frac{n}{2}(t - T) \tag{10.2}$$

whence

$$E - e \sin E = n(t - T).$$

The right-hand side of the equation is also called the *mean anomaly* M, and

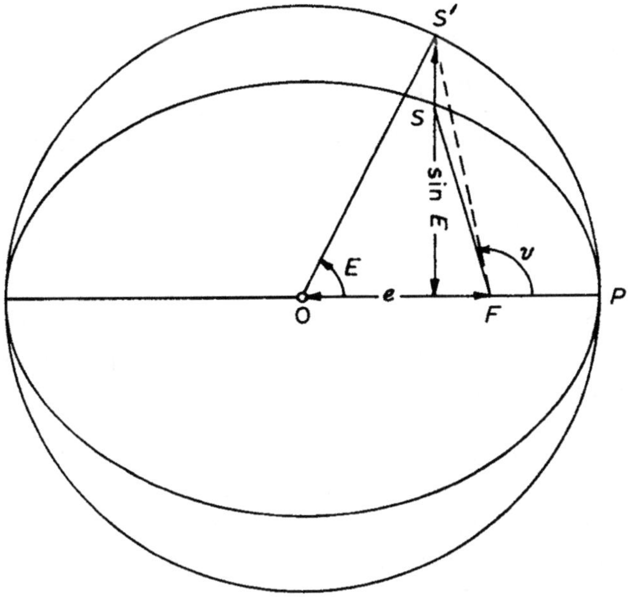

Fig. 10.3. Kepler's problem. Relation between position S on unit orbit and corresponding position S' on auxiliary unit circle.

thus we arrive at *Kepler's equation*

$$E - e \sin E = M, \tag{10.3}$$

a transcendental functional relation between E, e and M.

E as a function of M and e has been tabulated (Astrand, 1890).

To obtain the position in the orbit from E, we proceed as follows: υ may be calculated from (van de Kamp, 1967)

$$\tan \frac{\upsilon}{2} = \sqrt{\frac{1+e}{1-e}} \tan \frac{E}{2}, \tag{10.4}$$

and the radius vector r from

$$r = a(1 - e \cos E), \tag{10.5}$$

and thus the position in the orbit is known.

It is convenient to introduce the *elliptical rectangular coordinates* x, y, in the *unit orbit*, given by

$$\begin{aligned} x &= \cos E - e, \\ y &= \sin E \sqrt{1 - e^2}. \end{aligned} \tag{10.6}$$

Since

$$r \cos v = a(\cos E - e),$$
$$r \sin v = a \sin E \sqrt{1 - e^2}, \tag{10.7}$$

the elliptical rectangular coordinates in the orbit of semi-axis major a are thus given by

$$r \cos v = ax,$$
$$r \sin v = ay. \tag{10.8}$$

Tables exist for x and y as functions of e and M, so that E need not even be computed (Franz and Mintz, 1964). Current computer programs conveniently perform these calculations as the need occurs.

(d) *Apparent and true orbits. Orbital elements.* The determination of the orbit of a celestial object involves its orientation in space. The earth's orbital motion complicates the observed orbital motion of planets and comets. The latter complication is conspicuously absent from the *relative orbit* of the components of a binary star.

We shall limit ourselves to the astrometric approach to binaries, or the study of orbits as seen projected on the sky, i.e., perpendicular to the line of sight. We shall distinguish between *resolved* and *unresolved astrometric binaries*, depending on whether the components are observed separated or blended.

The relative orbit in its own plane is referred to as the *true orbit*, its orthogonal projection on the plane of the sky is the *apparent orbit*. The latter is obtained from observations of the relative positions obtained at different epochs; the first step is to draw the apparent orbit through the inherently imperfect observations in such a manner that it represents the projection of Keplerian motion. That is to say, the apparent ellipse has to obey the law of equal areas. Ideally, the observations should cover at least one full period of revolution. If not, an attempt is made to complete the apparent orbit on the basis of the observed part of the orbit.

To permit a more accurate drawing of the orbit it is generally desirable to make two diagrams – one of the position angle θ and one of the distance or separation ρ, both plotted against the time t, i.e., the epoch of observation. These diagrams permit an examination of the internal consistency and reveal extreme discordances of the observations. Normal places may be used by combining several observations. Except for very short periods the observations are usually combined into weighted means for each year. Visual and photographic observations are kept separate, and different weights are assigned to the corresponding normal points. For a general statement we may say that photographic normal places have about ten times the weight of visual normal places. In visual observations of double stars the distances are generally much less reliable than the position angles. Hence, whenever possible, the visual distances are not used, in combination with more recent photographic observations, only the position angles.

The projected law of areas requires that $\rho^2(\mathrm{d}\theta/\mathrm{d}t)$ = constant: $\mathrm{d}\theta/\mathrm{d}t$ may be expressed in radians per year; the quantity $\rho^2(\mathrm{d}\theta/\mathrm{d}t)$ is called the areal constant. By drawing smooth interpolation curves through the normal places the quantities ρ and $\mathrm{d}\theta/\mathrm{d}t$ can be read off for a number of epochs, and the constancy of $\rho^2(\mathrm{d}\theta/\mathrm{d}t)$ tested. Thus the reliability of the observations is revealed, and the interpolation curves may be adjusted to some extent, in order to yield more nearly equal values of $\rho^2(\mathrm{d}\theta/\mathrm{d}t)$ for different times. Adjusted interpolation curves are used to draw the apparent orbit, which most nearly satisfies the law of areas.

The true orbit in space of the companion relative to the primary is defined by seven elements that fall into three groups.

Dynamical elements. P = period, e = eccentricity, T = epoch of periastron passage. P and T are usually expressed in years. The mean motion in degrees per year is denoted by

$$n = \frac{360°}{P}.$$

Scale of orbit (i.e., semi-axis major or mean distance a, the average between periastron and apastron distances). In the orbital relationships, which follow, the symbol a generally refers to the (unforshortened) angular value, but the arcsecond symbol is omitted. Only when the parallax p is known, does conversion from angular to linear value become possible and significant (Chapter 12):

$$a(\mathrm{AU}) = \frac{a''}{p''}. \tag{10.9}$$

The dynamical elements P, e, and T, and the scale a determine the position in the true orbit. It is often useful to introduce the *unit orbit*, defined by the dynamical elements only, which is converted into the true orbit by applying the scale factor a.

Orientation elements (Figure 10.8) Ω represents the position angle of the ascending node – i.e., the point on the intersection of the orbital plane and the plane of the sky at the primary – at which the companion recedes from us. In the absence of appropriate radial-velocity observations the ascending node cannot be distinguished and Ω refers simply to that nodal point for which $\Omega < 180°$.

ω, sometimes referred to as 'longitude of periastron', is the angle in the plane of the true orbit from the ascending node (or nodal point) to periastron, in the direction of orbital motion.

i is the inclination of the orbital plane to the plane tangent to the sky. For direct motion, i.e., position angles increasing with the time, i is between 0° and 90°, for retrograde motion $90° < i < 180°$.

If $i = 0°$ or 180°, Ω is taken to be 0°, and the position angle of periastron is ω for direct ($i = 0°$), $360° - \omega$ for retrograde ($i = 180°$) motion.

If $e = 0$, take $\omega = 0°$, and T as the epoch of nodal passage.

It is logical and convenient to carry out the derivation of the true orbit from the apparent orbit in two steps. First, derive the dynamical elements, then derive the scale and the orientation elements, either by geometric or by analytical methods.

Scale and orientation elements are the 'conventional' *geometric elements*; these four elements may be transformed into other 'natural' geometric elements (Section f).

The designation *Campbell elements* is used to include the dynamical elements P, e, and T, the scale a, and the orientation elements Ω, ω, and i.

Projection of the Keplerian motion yields an elliptical motion for the secondary, which still obeys the law of areas with respect to the primary, although this is not at the focus of the apparent ellipse. The center of the apparent ellipse corresponds to the projected center of the true ellipse.

Generally no valid analysis of the orbit can be made till the part of the orbit that has been covered is sufficient to permit an acceptable estimate of

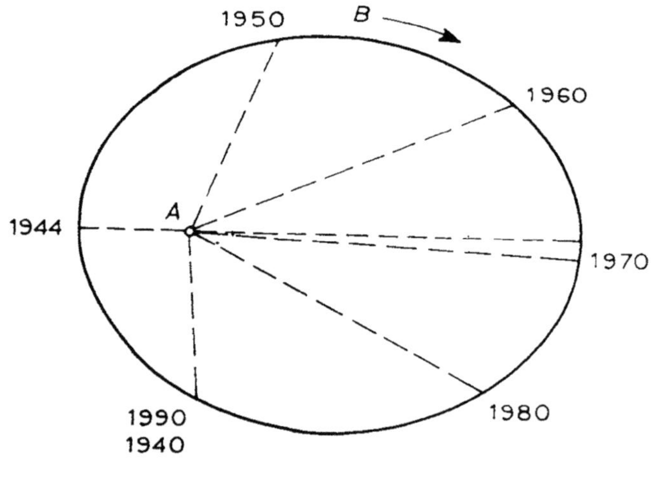

Fig. 10.4. Keplerian motion in true orbit for the visual binary Sirius.

the course of the apparent orbit through its complete revolution. The degree of completeness of the analysis depends on the apparent size of the orbit and the observational accuracy, and the time covered by the observations.

The adjusted apparent ellipse may now be analyzed and the orbital elements determined. The diameter of the apparent ellipse passing through the primary is the projected major axis of the true orbit; the extremes of this diameter are periastron and apastron. Any line parallel to the intersection of the true and apparent orbits, the so-called line of nodes, remains parallel and is not foreshortened in the apparent ellipse. Any line perpendicular to the line of nodes remains perpendicular, but its length is foreshortened in the apparent ellipse by the factor cos i. The auxiliary Kepler circle appears as an auxiliary Kepler ellipse (Section f) (Figure 10.7).

There are different ways in which to derive the orbital elements; we shall outline the natural and most frequently used methods of analysis. Most of them require first a determination of the dynamical elements, and we therefore consider this problem first.

(e) *Derivation of dynamical elements. From apparent orbit.* – The period P is measured from the graph in which the position angle θ is plotted against the epoch of observation. If more than one revolution has been completed, the derivation is quite straightforward. If less than one revolution has been completed, the interpolation curves for ρ and θ may be extrapolated, always trying to satisfy the relation $\rho^2(d\theta/dt)$ = constant. In this way the period may be estimated and the apparent ellipse is thus also tentatively completed.

The eccentricity e is measured from the apparent ellipse as the ratio of the

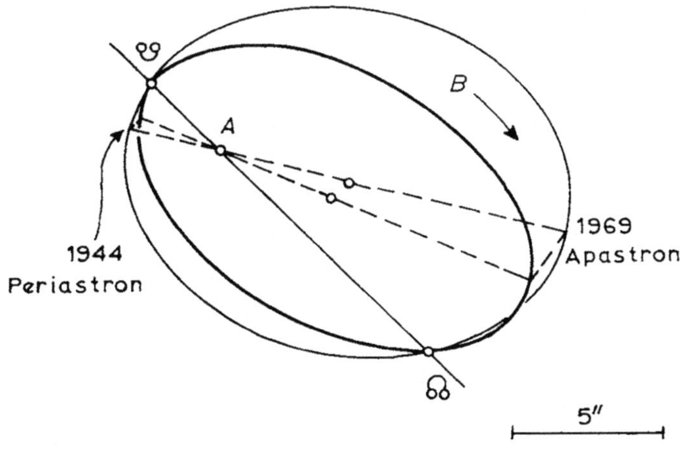

Fig. 10.5. Relation between true and apparent orbits of the visual binary Sirius.

lengths, primary-center and periastron-center; this ratio is not changed by projection.

The epoch of periastron passage T is obtained from the interpolation curve for θ by measuring the position angle of periastron passage as revealed by the projected major axis of the true ellipse.

The above procedure is inadequate in certain cases, particularly when the inclination of the orbit is near 90°: the apparent ellipse may then be very slender, or may appear as a straight line, so that the law of areas can no longer be used.

In these cases, and also for any general case, another approach may be used. Instead of studying the apparent ellipse and its implicit time element, it is equally (and often more) convenient and profitable to make use of the so-called time-displacement curves. (van de Kamp, 1947a).

From time-displacement curves. – Project the apparent orbital motion in any chosen direction, and plot this projection as a function of the time. Generally it is natural and convenient to study the projection of the apparent orbital motion in the two coordinates, right-ascension (x) and declination (y), and plot these projected motions against the time. This is completely analogous to the orbital study of spectroscopic binaries, where there is no other choice but to study the orbital motion in the one coordinate available to the spectroscopic observer, namely the line of sight. The visual-double-star observer has two coordinates at his disposal, as contrasted with the one coordinate available to the spectroscopic observer. The latter has one radial velocity curve; the visual observer has two displacement curves, whose combination gives the projection of the three-dimensional orbital motion on the two-dimensional plane of the sky.

The advantage of using displacements (and velocities) plotted against the time is that the best known observed datum, the time, enters explicitly. Periastron and apastron are located by the fact that their mean anomalies differ by 180° and their ordinates are equal and opposite when referred to the center of the orbit. Periastron and apastron are conveniently located by making a copy of the displacement curve, reversing it along the central line representing the center of the orbit and shifting the reversed curve half a period along the time axis. Generally two paires of intersections result (Fig 10.6).

Of these the single intersection on the shorter, steeper branch and the central intersection on the longer branch represent periastron and apastron, respectively.

The slopes of any displacement curve represent projected velocities dR/dt. The ratio of the true velocity vectors at periastron and apastron is $-(1 + e)/(1 - e)$, their directions being opposite. Since this ratio remains the same in projection, the ratio of the slopes $(\mathrm{d}R/\mathrm{d}t)_P$ and $(\mathrm{d}R/\mathrm{d}t)_A$ at periastron and apastron respectively amounts to $-(1+e)/(1-e)$, for any displacement curve, including those of astrometric orbits seen on edge. We thus cannot only distinguish periastron from apastron, but also derive the eccentricity of the orbit, independently of the focus, through the relation

$$e = \frac{(\mathrm{d}R/\mathrm{d}t)_P + (\mathrm{d}R/\mathrm{d}t)_A}{(\mathrm{d}R/\mathrm{d}t)_P - (\mathrm{d}R/\mathrm{d}t)_A}\,. \tag{10.10}$$

The method becomes unreliable when periastron and apastron are close to the extreme amplitudes – this occurs if the major axis is at a small angle with the line of nodes – for any spectroscopic orbit and for any astrometric orbit with a high inclination. In this case a combination of graphical and analytical mothods may be used to derive the eccentricity from the observed cosine of the eccentric anomaly for selected values of the mean anomaly (van de Kamp, 1947a). The method is particulary useful for 'linear' orbits seen nearly edgewise.

This analysis of time-displacement curves is equally applicable to resolved and to unresolved astrometric binaries, provided, of course, that in the latter the orbital motion of the center of light is appreciable and well separated from the proper motion and parallax of the system. Although the time-displacement curve method may often be superfluous for resolved binaries, it is of particular significance for unresolved binaries, for which the conventional geometric derivation of the eccentricity is awkward or fails, but the time-displacement curve method is effective, accurate, and generally applicable (Chapter 13).

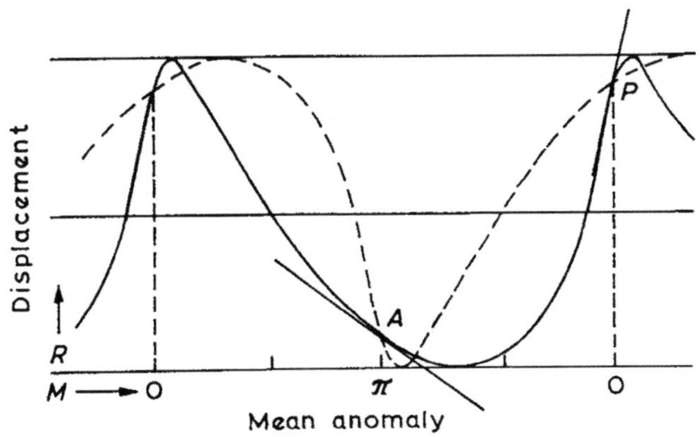

Fig. 10.6. Graphical evaluation of periastron P, apastron A, and eccentricity e from slope of time-displacement curve at periastron and apastron. The curve illustrated is $\cos E + 0.174 \sin E$, corresponding to $e = 0.7$.

For the sake of completeness we give the corresponding relation for analysis of the radial velocity pattern V/t of a spectroscopic binary:

$$e = \frac{\left(\frac{d^2R}{dt^2}\right)_P^{1/2} + \left(\frac{d^2R}{dt^2}\right)_A^{1/2}}{\left(\frac{d^2R}{dt^2}\right)_P^{1/2} - \left(\frac{d^2R}{dt^2}\right)_A^{1/2}}. \tag{10.11}$$

(f) *Derivation of geometric elements: Thiele-Innes constants.* After the dynamical elements have been obtained by either of the methods outlined above, the geometric elements may be determined by either geometric or analytic methods.

Geometric method. – This is often referred to as Zwiers' method (1895), although it was independently presented by Henry Norris Russell in a slightly different form (1898). We follow essentially Russell's presentation. The projected minor axis of the apparent orbit is obtained as the conjugate

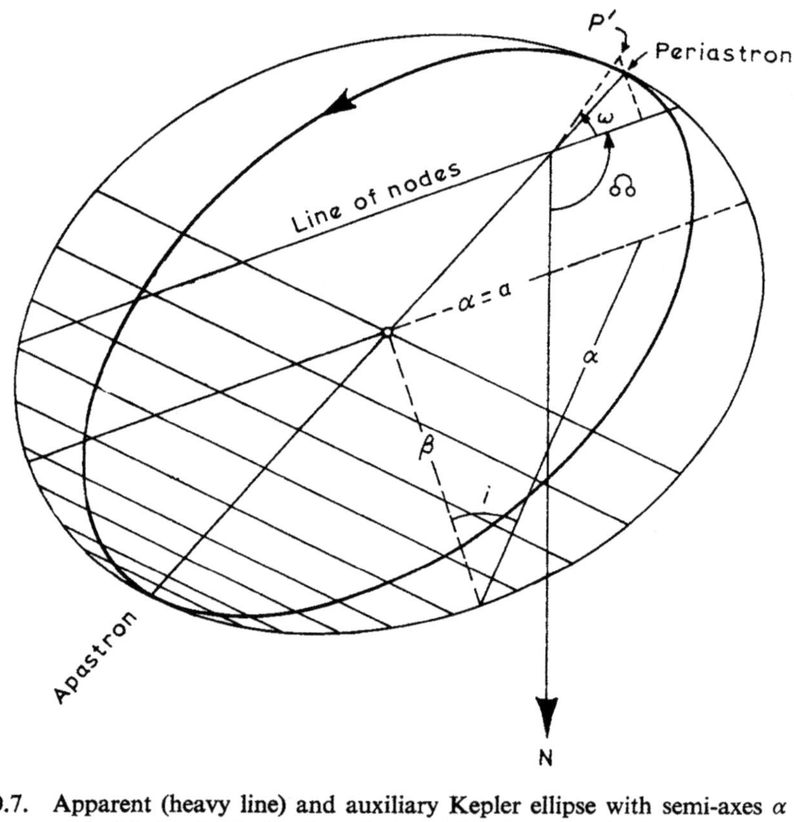

Fig. 10.7. Apparent (heavy line) and auxiliary Kepler ellipse with semi-axes α (major) and β (minor). Evaluation of conventional geometric elements: $e = 0.71$: $k = 1.40$; $i = 40°$; $\omega = 36°$; $☊ = 114°$.

diameter of the orthogonal projection of the true major axis. The projection of the auxiliary or Kepler circle has as conjugate diameters the projection of the major axis and the projection of the minor axis increased in the ratio

$$k = \frac{1}{\sqrt{1 - e^2}},$$

e having been previously measured. The projection of the *auxiliary* circle, may be readily constructed, it is called the *auxiliary ellipse* and has semi-axes α (major) and β (minor). It major axis is the only diameter of the auxiliary circle, which appears unforeshortened, and its length 2α is therefore equal in length, $2a$, to the major axis of the true orbit and parallel to the line of nodes, thus yielding the elements a and Ω. The cosine of the inclination i of the true orbit is given by the ratio of the minor to the major axes of the auxiliary ellipse, i.e., $\cos i = \beta/\alpha$, which may be constructed (Fig. 10.7).

All orbital distances perpendicular to the line of nodes are foreshortened by the factor $\cos i$. If we wish, we can therefore construct the true ellipse by applying the factor $1/\cos i$ perpendicular to the line of nodes to all points of the apparent orbit. In particular we may thus construct the unprojected position P' of the periastron in the true orbit, and hence measure the angle ω.

In practice this geometric method relies on analytical procedure also (Smart, 1960).

Analytic method. – An extremely elegant approach is the widely used Thiele-Innes method. Recall that for a position in the apparent orbit

$$\begin{aligned} \Delta\alpha \cos \delta &= \Delta X = \rho \sin \theta, \\ \Delta\delta &= \Delta Y = \rho \cos \theta. \end{aligned} \tag{10.1}$$

For the position in the true orbit

$$ax = r \cos \upsilon, \qquad ay = r \sin \upsilon. \tag{10.8}$$

The two sets of polar coordinates are related as follows (Figure 10.8):

$$\begin{aligned} AC &= \rho \cos (\theta - \Omega) = r \cos (\upsilon + \omega), \\ B'C &= \rho \sin (\theta - \Omega) = r \sin (\upsilon + \omega) \cos i. \end{aligned} \tag{10.12}$$

Working out these relations we find

$$\Delta X = Bx + Gy, \qquad \Delta Y = Ax + Fy \tag{10.13}$$

where

$$\begin{aligned} B &= a(\cos \omega \sin \Omega + \sin \omega \cos \Omega \cos i), \\ A &= a(\cos \omega \cos \Omega - \sin \omega \sin \Omega \cos i), \\ G &= a(-\sin \omega \sin \Omega + \cos \omega \cos \Omega \cos i), \\ F &= a(-\sin \omega \cos \Omega - \cos \omega \sin \Omega \cos i). \end{aligned} \tag{10.14}$$

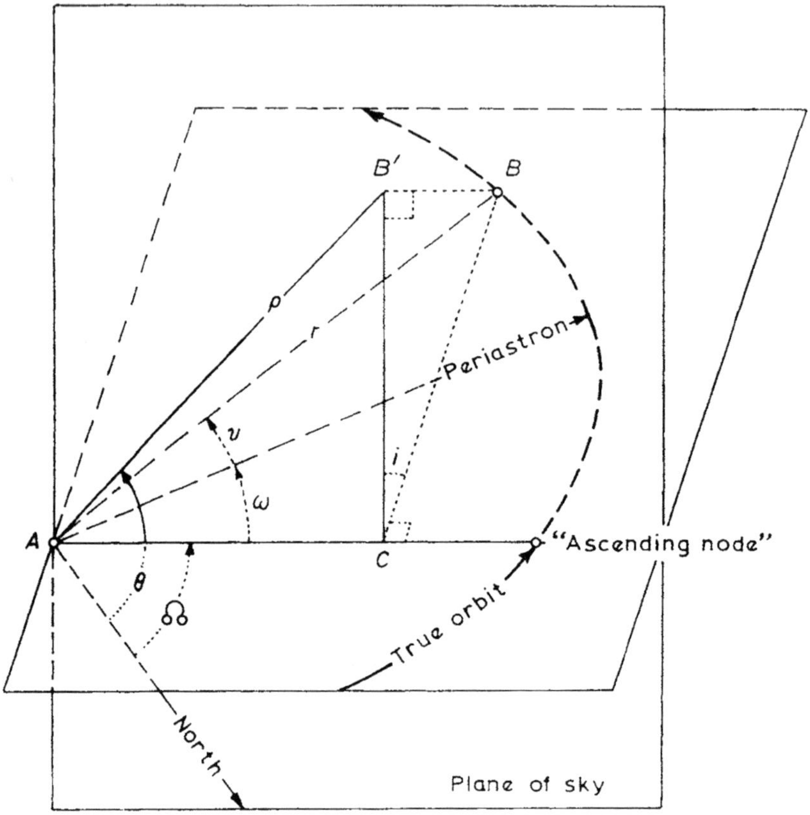

Fig. 10.8. Relation between polar coordinates in apparent and true orbits.

In these formulae x and y are the elliptical rectangular coordinates in the unit orbit (Section c): they are functions of the dynamical elements only; B, A, G, and F contain the scale a and the three orientation elements. These four *geometric* elements are also called *natural* elements, or Thiele-Innes constants, and are related to the conventional geometric elements through the relations (10.14). The designation Thiele-Innes elements includes both the dynamical elements P, e, and T and the Thiele-Innes constants B, A, G, and F.

The natural elements have a simple geometric meaning (Figure 10.9). The projections on any co-ordinate are given by $(\gamma_1 ax + \gamma_2 ay)$ in the apparent orbit and $(\gamma_1 a \cos E + \gamma_2 a \sin E)$ in the auxiliary ellipse. Here γ_1 and γ_2 are the direction cosines of the projected co-ordinates on the directions x and y in the true orbit. Note that $\gamma_1 a$ is the projected co-ordinate of periastron ($E = 0$), $\gamma_2 a$ the projected co-ordinate of the point $E = 90°$ on the auxiliary circle; both co-ordinates refer to the center of the orbit. Hence, in the equatorial co-ordinate system, the elements B, A and G, F represent the projec-

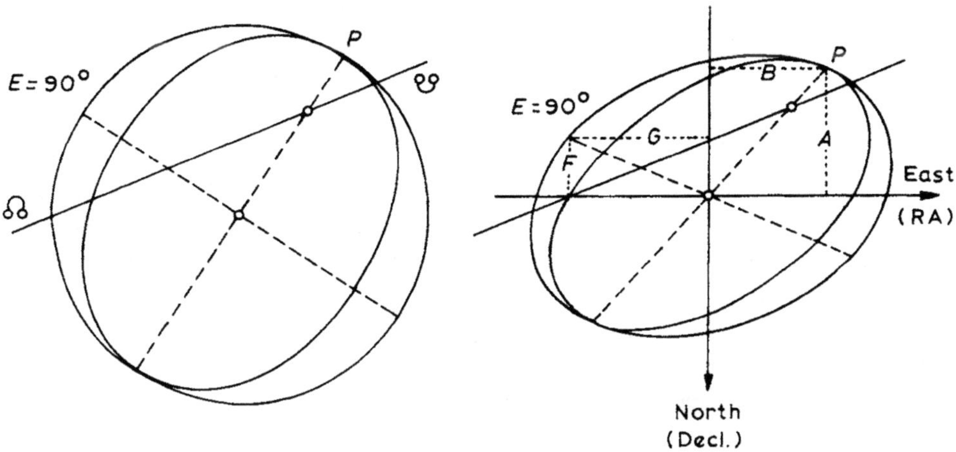

Fig. 10.9. (Left) True orbit and auxiliary Kepler circle. (Right) Apparent orbit and auxiliary Kepler ellipse in equatorial (rectangular) coordinate system. The Thiele-Innes constants are the projected rectangular equatorial coordinates of periastron and of the point $E = 90°$ on the auxiliary circle, referred to the center of the orbit.

tions of periastron and of the point $E = 90°$ on the auxiliary circle. Note that these quantities are proportional to the scale a, and refer to the center of the orbit.

(g) *Derivation of conventional from geometric elements.* The conventional geometric elements are derived as follows: relations (10.14) yield

$$\begin{aligned} A + G &= a(1 + \cos i) \cos (\omega + \Omega), \\ A - G &= a(1 - \cos i) \cos (\omega - \Omega), \\ B - F &= a(1 + \cos i) \sin (\omega + \Omega), \\ -B - F &= a(1 - \cos i) \sin (\omega - \Omega), \end{aligned} \tag{10.15}$$

hence we find ω and Ω from

$$\tan (\omega + \Omega) = \frac{B - F}{A + G},$$

where $\sin (\omega + \Omega)$ has the same sign as $B - F$, (10.16)

$$\tan (\omega - \Omega) = \frac{-B - F}{A - G},$$

where $\sin (\omega - \Omega)$ has the same sign as $-B - F$.

Hence the quadrants of $\omega + \Omega$ and $\omega - \Omega$ are unambiguously determined since $a(1 + \cos i)$ or $a(1 - \cos i)$ cannot be negative. However, we may add 360° to $\omega + \Omega$ (or to $\omega - \Omega$) and thus obtain two solutions

$$\omega, \Omega \quad \text{and} \quad \omega \pm 180°, \quad \Omega \pm 180°.$$

Unless a distinction is possible, the solution for which $\Omega < 180°$ is adopted.

To obtain a and i we apply the theorems of Apollonius for the ellipse.

(1) *The sum of the squares of any two conjugate diameters equals the sum of squares of the axes.*

(2) *The area between the parallelograms on any two conjugate diameters equals that of the rectangle of the axes.*

Applying these theorems to the auxiliary ellipse we find

$$a^2(1 + \cos^2 i) = A^2 + B^2 + F^2 + G^2 = 2k \qquad (10.17)$$

$$a^2 \cos i = AG - BF = m. \qquad (10.18)$$

We introduce the auxiliary quantities k and m and also

$$j^2 = k^2 - m^2 \qquad (10.19)$$

and find

$$a^2 = j + k \qquad (10.20)$$

$$\cos i = \frac{m}{a^2}. \qquad (10.21)$$

It has been found practical to determine the dynamical elements P, e, and T geometrically from photographic and visual normal places combined since generally the photographic normal places alone cover too short a part of the orbit. Next the geometric elements B, A, G, and F are determined from a least-squares solution based on the photographic normal places alone. The dynamical elements P, e, and T are given slight variations to test the stability of the solution and, if possible, to establish the set of elements that give the best fit, – as shown, for example, by a minimum value of the sum of the squares of the residuals, $O - C$. A simultaneous least-squares solution for corrections to all seven elements may be made if the available accuracy warrants such a procedure. Final residuals may be given both in x and y and in ρ and θ; the residuals in position angle may be stated in angle as well as in arc, i.e., $\rho_{\text{comp}} \sin(\theta_{\text{obs}} - \theta_{\text{comp}})$.

CHAPTER 11

PATH OF STAR WITH ORBITAL MOTION. PHOTOCENTER

A decade or so after the first determinations of stellar parallaxes by long-focus photographic astrometry, it became evident that by extending the observational material over a longer time interval, the same technique and methods could be used to measure the ratio of masses from the orbital motion in known visual binary systems having periods of the order of several decades. In this case an orbital term has to be added to the equations of condition for the analysis of the star's path. Orbital motions revealed by 'variable proper motion', i.e. perturbations, are studied in Chapter 13ff.

Generally, results for orbital motion in stellar paths measured photographically must be evaluated carefully, with particular attention given to separation of the components, which may vary widely over the interval studied. The simplest case is that of material with 'wide' separation of the components and/or large magnitude difference. In the latter case exposure and magnitude compensation with the reference stars may be adjusted to give a measurable image of the primary while the light contribution of the fainter secondary remains nil.

We distinguish between resolved and unresolved astrometric binaries.

(a) *Resolved astrometric binary; mass-ratio.* For a resolved astrometric binary, the path of the primary on a background of reference stars is represented by the equations of condition:

$$\begin{aligned} X_A &= c_X + \mu_X t + q_X t^2 + \pi P_\alpha - \mathrm{B}\varDelta X, \\ Y_A &= c_Y + \mu_Y t + q_Y t^2 + \pi P_\delta - \mathrm{B}\varDelta Y. \end{aligned} \tag{11.1}$$

Here c, μ, and q refer to the center of mass of the visual binary, B is the fractional mass of the secondary in terms of the combined mass of primary and secondary

$$\mathrm{B} = \frac{\mathcal{M}_B}{\mathcal{M}_A + \mathcal{M}_B} \,. \tag{11.2}$$

The scale a_1 of the orbit of the primary is B times that of relative orbit of the two components, i.e.

$$a_1 = \mathrm{B}a. \tag{11.3}$$

The difference in phase (position angle) is 180°; $\varDelta X$ and $\varDelta Y$ represent the relative position of the secondary B referred to the primary A, i.e. $X_B - X_A$ and $Y_B - Y_A$ respectively.

Similar equations, replacing $-\mathrm{B}$ by $(1 - \mathrm{B})$, hold for the secondary if the latter is recorded on the photographic plate and well separated from the primary; the scale of its orbit is $a_2 = (1 - \mathrm{B})a$ and the equations of condition become:

$$\begin{aligned} X_B &= c_X + \mu_X t + q_X t^2 + \pi P_\alpha + (1 - \mathrm{B})\Delta X, \\ Y_B &= c_Y + \mu_Y t + q_Y t^2 + \pi P_\delta + (1 - \mathrm{B})\Delta Y. \end{aligned} \tag{11.4}$$

To successfully determine B or $1 - \mathrm{B}$, the observational material has to be sufficiently strong and temporally extended to separate orbital motion from proper motion and quadratic time effect. Each case has to be considered on its own merits; a rule of thumb for minimum requirement is some 50 plates and a 'curvature' of 0."5 in the apparent relative orbit of the two components (van de Kamp, 1943).

The quadratic time effect may (have to) be eliminated by correcting for reduction to absolute and perspective acceleration (Chapter 8, Section g).

If both components are recorded and well separated on the photographic plate, ΔX and ΔY are known and only one set of equations obviously need be used. If ΔX and ΔY are available from an independently well established visual orbit, both sets of Equations (11.1) and (11.4) may be used and independent solutions made for B and $1 - \mathrm{B}$. If the secondary is too faint to be recorded and sufficiently separated from the primary to eliminate any possible blending effect equations (11.1) may be used if ΔX and ΔY are known from an established visual orbit.

A series of photovisual photographs (Figure 11.1) shows the proper motion and orbital motion of the components A and B of the visual binary Krüger 60, referred to the optical companion C, from exposures made with the McCormick 66-cm refractor (1919 and 1933) and with the Sproul 61-cm refractor (1938–1965). The exposures have been oriented so that the proper motion of the barycenter of A and B is horizontal; they have been shifted to align the exposures of C vertically. The vertical locations are a linear function of the epoch of observation and result in a rectilinear arrangement of the barycenter of A, B for the successive exposures.

The proper motion of the barycenter is

$$\mu = 0''.90 \quad \text{in} \quad \theta = 245^\circ$$

or

$$\mu_\alpha \cos\delta = -0''.81, \qquad \mu_\delta = -0''.38.$$

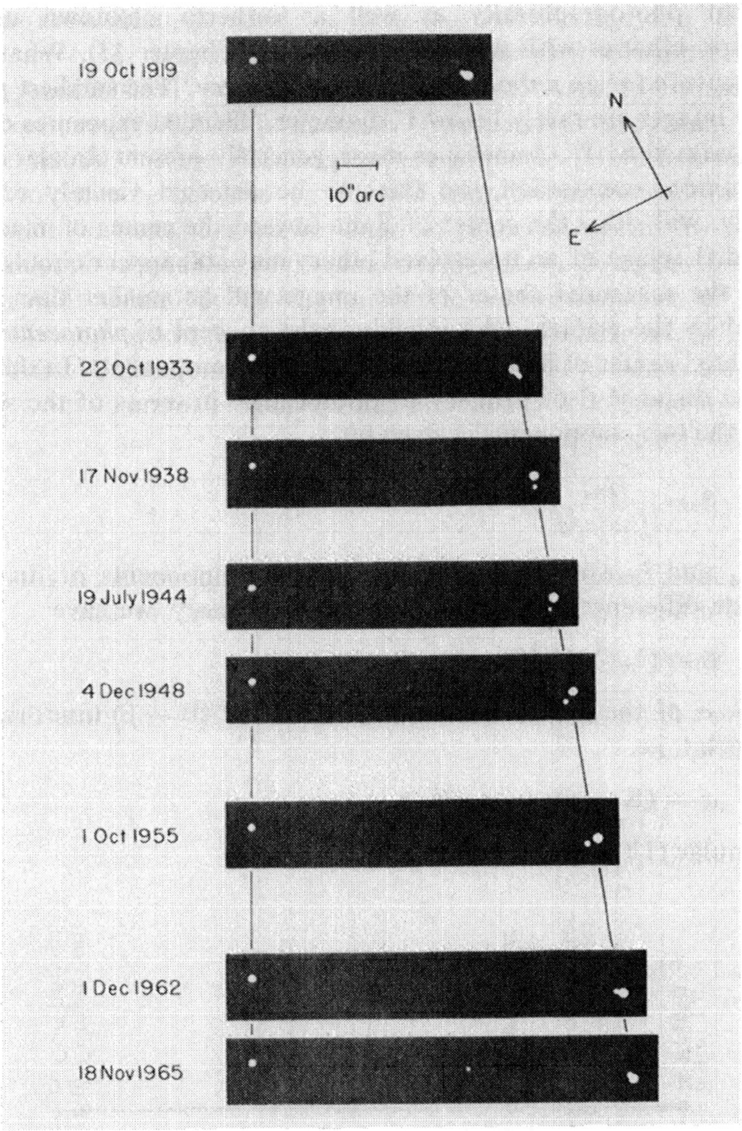

Fig. 11.1. Proper motion and orbital motion of the components *A* and *B* of the visual binary Krüger 60 relative the optical companion *C*.

The summer exposure (1944) shows a morning parallactic displacement of nearly 0″.4, relative to the other, evening exposures (Wanner, 1967).

(b) *Unresolved astrometric binary; photocenter and photocentric orbit.* In the case of photographic blending for close visual binaries formulae (11.1) require adjustment. This is the case for known binaries resolved visually but unresolved photographically as well as hitherto unknown unresolved astrometric binaries with unseen companions (Chapter 13). What position do we measure for an *unresolved astrometric binary*? The smallest photographic star images are rarely below 1″ diameter. Blended exposures of components separated by 1″, sometimes more, generally present circular images. A close luminous companion, too faint to be detected visually or spectroscopically, will draw the center of light toward the center of mass. Hence the blended image of an unresolved binary may still appear circular, but the orbit of the measured center of the image will be smaller than the orbit described by the primary. We introduce the concept of *photocenter*, ideally the weighted center of light intensity of the two components. In this case the fractional distance β of primary to photocenter in terms of the separation between the two components is given by

$$\beta = \frac{l_B}{l_A + l_B}, \tag{11.5}$$

where l_A and l_B are the luminosities of the components or, in terms of magnitude difference Δm, companion minus primary, we have

$$\beta = (1 + 10^{0.4\Delta m})^{-1}. \tag{11.6}$$

The *scale* α of the *photocentric orbit* is therefore $(B - \beta)$ times that of the relative orbit, i.e.

$$\alpha = (B - \beta)a \tag{11.7}$$

and formulae (11.1) are changed to

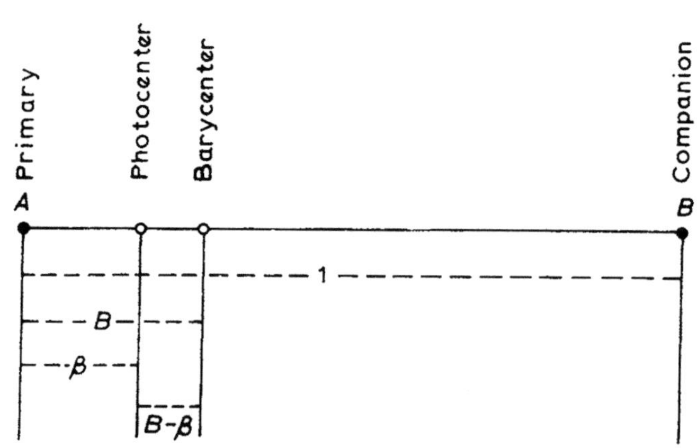

Fig. 11.2. Relative spacing between components, barycenter, and photocenter for photographically unresolved binaries.

$$X = c_X + \mu_X t + q_X t^2 + \pi P_\alpha - (\mathrm{B} - \beta)\Delta X,$$
$$Y = c_Y + \mu_Y t + q_Y t^2 + \pi P_\delta - (\mathrm{B} - \beta)\Delta Y. \tag{11.8}$$

Various investigators have found deviations for the observed values of β from the simple theoretical relation (11.5) (Hall, 1951; Feierman, 1971; Morel, 1970). The discrepancies depend on telescope, emulsion, separation, and other factors. For the Sproul refractor the observed values of β agree with the theoretical values up to $\Delta m = 1$. For larger magnitude differences and for separations generally less than 1″ (0.05 mm), the observed values of β are smaller than the theoretical values by approximately 0.05, or even by larger amounts for separations over 1″. In other words, for values of $\Delta m > 1$, the influence of the companion on the location of the photocenter is less than what follows from the theoretical relation. Ideally, as suggested by Feierman, if β could be evaluated as a function of the changing separation, the observed positions could be corrected and the following formulae used for the determination of B:

$$X - \beta\Delta X = c_X + \mu_X t + q_X t^2 + \pi P_\alpha - \mathrm{B}\Delta X$$

and (11.9)

$$Y - \beta\Delta Y = c_Y + \mu_Y t + q_Y t^2 + \pi P_\delta - \mathrm{B}\Delta Y.$$

(c) *Alternate analysis: parallactic and apparent orbit. Orbital factors.* In statistical astronomy each stellar proper motion is the resultant of the star's 'peculiar' and its (secular) parallactic motion. Similarly for a binary star each component has two periodic orbits, its *apparent* or 'own' orbit and the annual *parallactic* orbit, which is the reflection of the Earth's orbit around the Sun. Of the latter we know *a priori* all elements except the angular unforshortened semi–axis major i.e. the relative parallax π of the star.

Alternate formulae may now be used if the binary's orbital elements are known from an analysis covering (a sufficient part of) the relative orbit of the two components (van de Kamp, 1945). It is often more explicit and elegant to use the following homogenous equations of condition in which apparent and parallactic orbit appear, side by side, for the observed path of primary component or of photocenter:

$$X = c_X + \mu_X t + q_X t^2 + \pi P_\alpha + \alpha Q_\alpha,$$
$$Y = c_Y + \mu_Y t + q_Y t^2 + \pi P_\delta + \alpha Q_\delta. \tag{11.10}$$

The last two terms in (11.10) represent the parallactic and the orbital displacements in the respective coordinates. The annual parallax factors P_α, P_δ represent the projected fractional coefficients in RA and Decl. of the unforshortened angular value (relative annual parallax) of one astronomical unit (Chapter 5). The *orbital factors* Q_α, Q_δ are the corresponding projected fractional coefficients of the unforshortened angular value of the semi-axis major

$$\alpha = a_1 = \mathrm{B}a \quad \text{for primary (no blending),} \tag{11.3}$$

$$\alpha = (\mathrm{B} - \beta)a \quad \text{for photocenter.} \tag{11.7}$$

The orbital factors are the projected values in right ascension (reduced to great circle) and in declination of the radius vector barycenter – primary, or photocenter, for unit orbit. The orbital factors are analogous to the parallax factors; the latter refer to the star's parallactic orbit, te former to the star's own apparent orbit.

The orbital factors are calculated from (the) known dynamical and orientational elements of the relative orbit:

$$\begin{aligned} Q_\alpha &= (b)x + (g)y, \\ Q_\delta &= (a)x + (f)y. \end{aligned} \tag{11.11}$$

The *orientation* factors

$$\begin{aligned} (b) &= -\cos\omega\sin\Omega - \sin\omega\cos\Omega\cos i, \\ (a) &= -\cos\omega\cos\Omega + \sin\omega\sin\Omega\cos i, \\ (g) &= +\sin\omega\sin\Omega - \cos\omega\cos\Omega\cos i, \\ (f) &= +\sin\omega\cos\Omega + \cos\omega\sin\Omega\cos i. \end{aligned} \tag{11.12}$$

are functions of the orientation elements ω, Ω, and i of the relative orbit of the two components, while x and y are the elliptical rectangular coordinates in unit orbit (Chapter 10). The orientation factors are related to the Thiele-Innes constants as follows:

$$\begin{aligned} B &= -(b)a, \\ A &= -(a)a, \\ G &= -(g)a, \\ F &= -(f)a. \end{aligned} \tag{11.13}$$

where a is the scale (semi–axis major) of the relative orbit of primary and companion.

The sign of α is determined by the sign of (B $-$ β). For visual binaries on the main sequence (Feierman, 1971) B ranges from 0.32 to 0.49, with values of β ranging from 0 to 0.116, i.e., B $-$ β is always positive. For a main sequence star accompanied by a white dwarf, or a similar object of appreciable mass but little or no luminosity, β is virtually zero and again $B - \beta$ is positive. $B - \beta$ can be negative, that is β can exeed B in absolute value, only for a comparatively small magnitude difference and a very small fractional mass for the fainter component, a condition contrary to present knowledge of masses and luminosities, but in principle not excluded.

As mentioned earlier, in order to obtain a meaningful value of B or α, depending on which formulae we use, the observational material has to cover a portion of the orbit with sufficient orbital motion ('curvature') to separate it from the proper motion, and if necessary, from the quadratic time effect, of the barycenter (van de Kamp, 1943).

CHAPTER 12

MASS-RATIO AND MASSES. HARMONIC RELATION

(a) *Fractional mass, mass-ratio. Harmonic relation.* The fractional mass $B = \mathcal{M}_B/(\mathcal{M}_A + \mathcal{M}_B)$, and hence the mass ratio $\mathcal{M}_B/\mathcal{M}_A$ are thus obtained from observational material covering sufficient orbital motion (curvature), to separate it from the proper motion and if necessary from the quadratic time effect of the barycenter.

The fractional mass B for a resolved binary is found either directly, using (11.1), (11.2) or from the relation

$$B = \frac{a_1}{a} \tag{11.3}$$

if we use Equation (11.10) in which $\alpha = a_1$. For the case of blending, $B - \beta$ is found from (11.8) and corrected for β, to obtain B, or, again for constant β, from (11.10) which leads to

$$B = \frac{\alpha}{a} + \beta. \tag{11.7}$$

For the case of a changing β, which can be evaluated, B may be found from (11.9).

Relation (11.7) explicity reveals the accuracy of B as it depends on errors in α, a and β; the latter is often the principal source of error. This equation is of particular significance in the analysis of unresolved objects not previously known or recognized as binaries (Chapter 13). The mass-ratio of the two components A and B is thus obtained in the form of the fractional mass

$$B = \frac{\mathcal{M}_B}{\mathcal{M}_A + \mathcal{M}_B} \tag{11.2}$$

of the secondary in terms of the combined mass of primary and secondary. Or, we may write

$$\frac{\mathcal{M}_B}{\mathcal{M}_A} = \frac{B}{1 - B} \tag{11.2a}$$

Individual masses may then be found if the combined mass can be found with adequate accuracy. This may be done for a visual binary with well determined orbit and parallax through the *harmonic* mass – space – time relation:

$$\mathscr{M}_A + \mathscr{M}_B = \frac{a^3}{P^2} \,. \tag{12.1}$$

Here the units of mass, distance and time are the Sun's mass, astronomical unit of distance and the sidereal year. The angular value of the unprojected semi-axis major a'' of the relative orbit of B and A obtained from orbital analysis, (Chapter 10), may be converted to linear measure if the parallax p'' is known, by the relation

$$a = \frac{a''}{p''} \,. \tag{10.9}$$

The harmonic relation (12.2) thus becomes

$$\mathscr{M}_A + \mathscr{M}_B = \frac{a^3}{p^3}\,\frac{1}{P^2}, \tag{12.2}$$

where a and p are expressed in seconds of arc. At present the accuracy of the combined mass is primarily limited by the attainable accuracy in the parallax p, and for close binaries, in the value of a. Frequently the mass ratios are much better known than the masses because of the triple effect of errors in p and a. Accurate masses have been determined for only a few dozen visual binaries.

(b) *Mass-luminosity relation.* Astrometric results have contributed to our knowledge of the mass-luminosity relation in our neighborhood for stars with masses less than about twice the Sun's mass. The past decades have witnessed continued advances in our knowledge of the fainter portion of the mass-luminosity relation. Whereas before 1955 the smallest known accurate mass of a visible star was that of Krüger 60B ($0.16\mathscr{M}_\odot$) we have since found Ross 614A ($0.114 \pm 0.016\,\mathscr{M}_\odot$) and Ross 614B ($0.062 \pm 0.009\,\mathscr{M}_\odot$) (Lippincott and Hershey, 1972). The mass of the latter star is within the errors, both observational and theoretical, of the lower mass limit, calculated for the main sequence (Grossman, 1970; Straka, 1971).

Figure 12.1 shows the relation between mass and the luminosity (log mass vs absolute visual magnitude M_v) for visual binaries nearer than ten parsec. The data were kindly provided by H. Jahreisz from a variety of sources. The diagram clearly demonstrates the general relation between mass and luminosity which resembles closely, in fact is the (zero age) main sequence; striking exceptions are the white dwarfs Sirius B and Procyon B to the left of the main sequence. There are other, smaller deviations indicating components which are either over- or under-luminous for their mass.

Our knowledge of larger masses for more luminous stars depends mostly on spectroscopic and on eclipsing binaries.

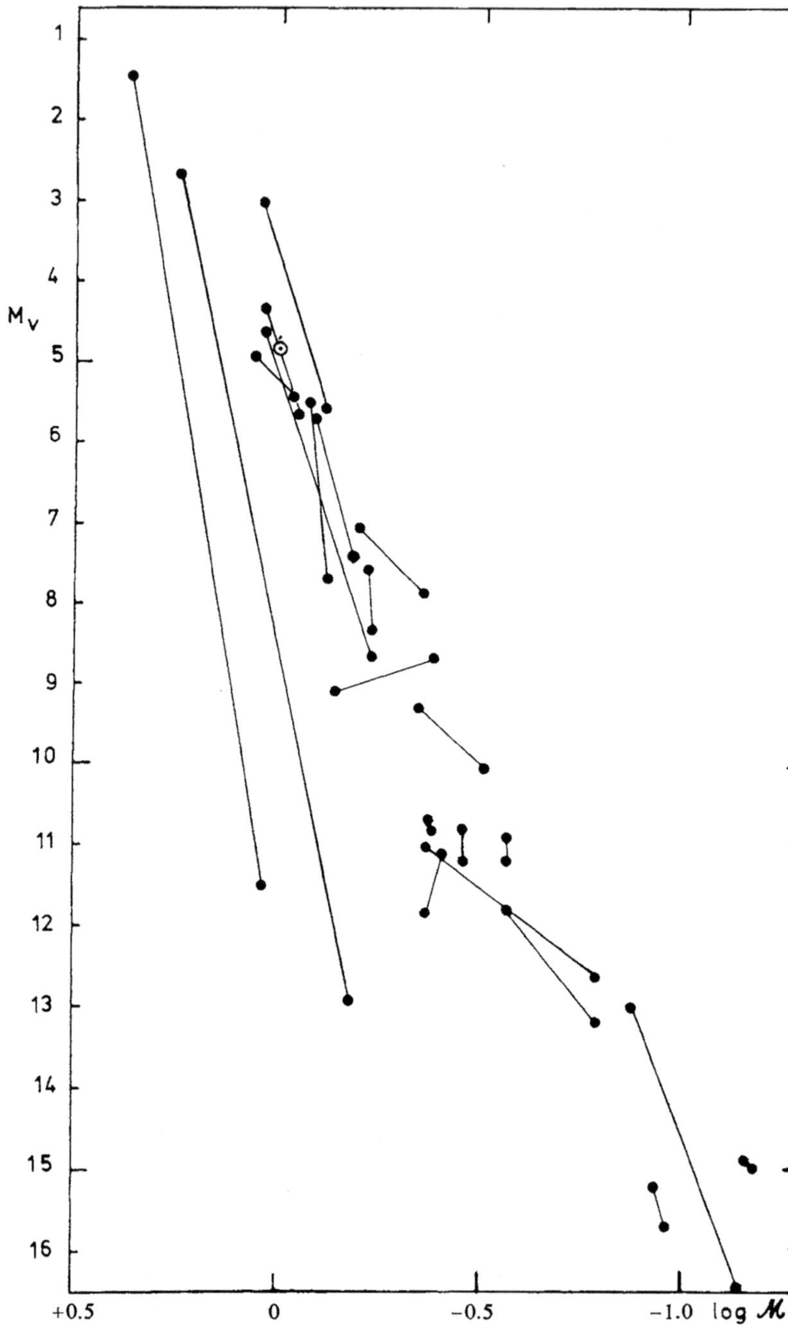

Fig. 12.1. Mass-luminosity relation for binary components nearer than 10 parsec. The Sun is indicated by ⊙.

A very precise mass-ratio and mass determination is found for Alpha Centauri AB, ($P = 79.92$ yr, $a = 17''.567$, $p = 0''.753 \pm 0''.004$). Kamper and Wesselink (1978) find

$$\frac{\mathcal{M}_B}{\mathcal{M}_A + \mathcal{M}_B} = 0.452 \pm 0.002$$

using both Yale (Johannesburg) and Cape plates over the interval 1926–1971. The resulting masses are

$$\mathcal{M}_A = 1.10\,\mathcal{M}_\odot, \qquad \mathcal{M}_B = 0.91\,\mathcal{M}_\odot$$

(c) *Mass-ratio determination for a long-period visual binary. Example: 61 Cygni.* For binary stars of very long period (several centuries), with well separated, measurable, components, observed over an interval small compared with the period, we may proceed in a different manner to obtain a value for the mass-ratio. The orbital motion, to a high degree of approximation, may be represented as a quadratic time effect. The orbital effects on the two components are of opposite sign, differing 180° in position angle, and are in inverse ratio to their masses. Assume that we are in a position to calculate both the reduction to absolute and the perspective acceleration (Chapters 6 and 8).

We now represent the measured positions of both components by the formulae for the path of a single star:

$$\begin{aligned} X_A &= c_{XA} + \mu_{XA}t + q_{XA}t^2 + \pi P_\alpha, \\ Y_A &= c_{YA} + \mu_{YA}t + q_{YA}t^2 + \pi P_\delta, \\ X_B &= c_{XB} + \mu_{XB}t + q_{XB}t^2 + \pi P_\alpha, \\ Y_B &= c_{YB} + \mu_{YB}t + q_{YB}t^2 + \pi P_\delta. \end{aligned} \tag{12.3}$$

Least squares solutions over the interval of observations yield values for the quadratic term coefficients, assuming that these are well separated from the proper motion terms. Each quadratic time coefficient is the sum of the coefficient of the orbital effect of the component relative to the barycenter, and the quadratic time coefficient of the barycenter, which is the sum of the reduction to absolute (Chapter 6) and the perspective acceleration (Chapter 8). The former may be calculated if the proper motions of the reference stars are determined, the latter follows from the known values of proper motion, radial velocity and parallax of the (barycenter of the) binary system. The observed quadratic coefficients are corrected to the barycenter as follows:

$$\begin{aligned} q_{XAc} &= q_{XA} + [\Delta D\mu_x] + 1''025 \times 10^{-6}\, \mu_X V p \\ q_{YAc} &= q_{YA} + [\Delta D\mu_y] + 1''025 \times 10^{-6}\, \mu_Y V p \\ q_{XBc} &= q_{XB} + [\Delta D\mu_x] + 1''025 \times 10^{-6}\, \mu_X V p \\ q_{YBc} &= q_{YB} + [\Delta D\mu_y] + 1''025 \times 10^{-6}\, \mu_Y V p \end{aligned} \tag{12.4}$$

These corrected values q_c represent the curvature of orbital motion, and lead to values of the mass-ratio:

$$\frac{\mathcal{M}_B}{\mathcal{M}_A} = -\frac{q_{Ac}}{q_{Bc}} \tag{12.5}$$

both in x and y.

An example is the long-period classical binary 61 Cygni ($P = 720$ yr, $a = 24''575$). A total of 966 plates taken with the Sproul 61-cm refractor on 280 nights over the interval 1912–1972 were measured on the two-coordinate Grant machine of Sproul Observatory (van de Kamp, 1973). Four reference stars were used; their relative motions were determined on a background of fifteen stars. Parallax, proper motion and radial velocity of this system are well known:

$$p = 0''292, \quad \mu_X = +4''13, \quad \mu_Y = +3''20, \quad V = -64 \text{ km s}^{-1}.$$

Least squares solutions of Equations (12.3) yield values for q, given in Table 12.1, together with the required corrections (12.4), to obtain the corrected values q_c.

TABLE 12.1

Yearly quadratic time effects for 61 Cygni A and B. Unit 0''000001

	x				y			
	A		B		A		B	
q	+218	± 4	− 67	± 5	−114	±4	+191	±5
reduction to absolute	− 3	12	− 3	12	+ 20	7	+ 20	7
correction for perspective	− 79		− 79		− 61		− 61	
q_c	+136	±12	−149	±12	−155	±8	+150	±8

The principal source of error in q_c lies in the error of the reduction to absolute, due to the cosmic errors or dispersion in the proper motions of the available background stars.

The mass ratio is found to be

$$\frac{\mathcal{M}_B}{\mathcal{M}_A} = \begin{matrix} 0.91 \pm 0.08 & \text{RA} \\ 1.03 \pm 0.05 & \text{Decl.} \end{matrix}$$

and we adopt

$$\frac{\mathcal{M}_B}{\mathcal{M}_A} = 0.99 \pm 0.04 \quad \text{giving double weight to Decl.}$$

This provisional result for the mass-ratio of 61 Cygni is consistent with what might be expected for the mass-ratio of two components with absolute visual magnitudes 7.5, 8.3 and spectral types K5, K7.

(d) *Derivation of harmonic relation.* The dynamical situation for the relative orbit of two masses is expressed by Newton's law of gravitation

$$f = -G\frac{\mathcal{M}_A + \mathcal{M}_B}{r^2}, \tag{12.6}$$

where f is the mutual acceleration along the radius vector r, and G the constant of gravitation. 6.67×10^{-8} in the c.g.s. system.

The kinematical statement in polar coordinates for this central (rather 'focal') acceleration is

$$f = \frac{\mathrm{d}^2 r}{\mathrm{d}t^2} - r\left(\frac{\mathrm{d}\upsilon}{\mathrm{d}t}\right)^2. \tag{12.7}$$

For an elliptical orbit (Kepler's first law) with parameter (half latus rectum) p, eccentricity e and true anomaly υ), we have

$$r = \frac{p}{1 + e\cos\upsilon} \tag{12.8}$$

from which:

$$\frac{\mathrm{d}r}{\mathrm{d}t} = \frac{e}{p} r^2 \sin\upsilon \frac{\mathrm{d}\upsilon}{\mathrm{d}t} = \frac{2Ae}{p}\sin\upsilon, \tag{12.9}$$

where

$$A = \frac{r^2}{2}\frac{\mathrm{d}\upsilon}{\mathrm{d}t} \tag{12.10}$$

is the constant areal velocity (Kepler's second law).

Hence the first term of (12.7) is

$$\frac{\mathrm{d}^2 r}{\mathrm{d}t^2} = \frac{2Ae}{p}\cos\upsilon\frac{\mathrm{d}\upsilon}{\mathrm{d}t} = \frac{4A^2}{pr^2}e\cos\upsilon \tag{12.11}$$

and the second term is

$$-r\left(\frac{\mathrm{d}\upsilon}{\mathrm{d}t}\right)^2 = -\frac{4A^2}{r^3}. \tag{12.12}$$

Substitution of (12.11) and (12.12) in (12.7) yields

$$f = \frac{4A^2}{r^2}\left(\frac{e\cos\upsilon}{p} - \frac{1}{r}\right) = -\frac{4A^2}{pr^2}. \tag{12.13}$$

The areal velocity in terms of the space-time dimensions of the elliptic orbit is

$$A = \frac{\pi a^2\sqrt{1-e^2}}{P} \tag{12.14}$$

whence

$$A^2 = \pi^2 p\frac{a^3}{P^2}, \tag{12.15}$$

where a is the semi-axis major and P the period. The eccentricity is now eliminated since $p = a(1 - e^2)$.

Substitution in (12.13) yields

$$f = -4\pi^2\frac{a^3}{P^2}\frac{1}{r^2}. \tag{12.16}$$

Equating the dynamical (12.6) and kinematical (12.16) statements we find

$$\mathcal{M}_A + \mathcal{M}_B = \frac{4\pi^2}{G}\frac{a^3}{P^2} \tag{12.17}$$

for the combined mass, using c.g.s. units.

In terms of the units Sun's mass, astronomical unit of distance and sidereal year, the constant of gravitation becomes

$$G = 4\pi^2 \tag{12.18}$$

neglecting the mass of the Earth and we obtain the simple harmonic relation

$$\mathcal{M}_A + \mathcal{M}_B = \frac{a^3}{P^2} \tag{12.1}$$

which is a generalization of Kepler's third or harmonic law.

CHAPTER 13

PERTURBATIONS IN STELLAR PATHS. HISTORY. ANALYSIS

(a) *History, Discovery.* Long-focus photographic astrometry has proven suitable and useful for the discovery and study of perturbations, revealed either as 'variable' proper motion of a star previously considered single, or as an irregularity in the Keplerian motion of a visual binary. In either case unseen companions are indicated; the method is particularly effective for large differences in magnitude between primary and companion.

The astrometric technique is valuable for nearby stars, the great majority of which are of faint apparent magnitude. The spectroscopic approach remains important for apparently bright stars. The classical first, unplanned photographic discovery of a perturbation was that of the red dwarf Ross 614, from only 25 plates spread over nine years, at the McCormick Observatory by Reuyl (1936). The companion was first seen and photographed by Walter Baade, close to greatest separation in 1955, and was measured visually by several others later. The early history of this object has been told elsewhere (Lippincott, 1955a, b). Another interesting case is the photographic perturbation of VW Cephei, discovered, and the companion seen afterwards, at the Sproul Observatory (Hershey, 1975a). For both stars the stellar companions were seen at the predicted positions. Further details about Ross 614 and VW Cephei are given in Chapter 14.

More than twenty well-established perturbations are now known, several others are provisional or uncertain. The great majority of the unseen companions prove to be *bona fide* stars with masses above $0.06\mathscr{M}_\odot$, on or close to the main sequence. One striking exception is the still unseen massive companion ($0.9\mathscr{M}_\odot$) of Zeta Cancri C, most likely a white dwarf (van de Kamp, 1947b). Objects below the critical mass of $0.06\mathscr{M}_\odot$ include the unseen companions of BD + 68° 946, BD + 43° 4305, Stein 2051 A, CC 1228 and the planet-like companion(s) of Barnard's star about which more will be said in Chapter 16.

The discovery of the perturbation of Ross 614 as well as that of several other stars was not the result of a planned program, but was a by-product of conventional parallax studies. Systematic searches were begun in the early 1930's at the Leander McCormick Observatory and have been carried out intensively since 1937 at the Sproul Observatory, where all stars within 10 pc, sometimes beyond, and within reach of the 61-cm refractor are photographed on a regular basis, preferably each observing season and, for the very nearest stars on several nights each year. The majority of stars on this program are nearby late-type dwarf stars; because of their relatively small masses the chances of finding a perturbation are much more favorable than they are for solar-type stars, though some of these have revealed perturbations also. Several other observatories have contributed to this field of research, notably Allegheny, McCormick, and increasingly so the United States Naval Observatory.

Thousands of parallax determinations of 'single' stars have been made from material spread over a few years. Perturbations of 'short' periods, of the order of a year or less, generally have small amplitudes and are not easily found. With few exceptions it usually has been possible to represent a short-time path satisfactorily by uniform proper motion and an annual parallactic orbit; moreover no increase in the probable error of one plate has been found for stars of large parallax. Perturbations of 'long' periods of the order of several decades, can not reveal any measurable perturbation in a few years; continued observations are required. Orbital motion may be partly absorbed in proper and parallactic motion, or may escape detection because of temporal gaps in the series of observations.

Narrow hour-angle requirements cause annual gaps of six to seven months, which may result in spurious periods: positions in successive cycles of a short-period orbit may be interpreted by a multiple, or a fraction of the true period. As long as the binary is not resolved, the analysis lacks the control of the harmonic relation (12.1), which for resolved binaries serves as a guide for the period.

Interpreting limited material by a fortuitous orbit may be avoided through continued observations. The multiple-exposure technique (Chapter 10) is ideally suited for the discovery and study of perturbations in well-resolved visual binaries. For close binaries visual and speckle techniques remain a desirable observational approach.

A satisfactory perturbation orbit generally is not obtained until all phases of the orbit have been covered. Correct dynamical interpretation is aided by the fact that, as a rule, Keplerian motion is observed in both RA and Decl. We must also reckon with the possibility that the observed perturbation is the result of two (or more) perturbations caused by two (or more) companions. Generally however, at the present stage of the art, one may have to be satisfied with finding a perturbation due to one, at most two comparatively massive companions of either stellar, substellar or Jupiter-like mass. Perturbations due to Earth like planets are much too small to be found, which of course is no argument against the possible presence of such objects.

For dynamical interpretation of a perturbation, the two important orbital element are the *semi-axis major* α and the *period* P.

(b) *Orbital analysis: dynamical and geometric elements.* In what follows some similarities with portions of Chapter 10 to which the reader is referred, unavoidably appear.

The remainders R from a solution for proper motion and parallax (and quadratic time effect, if need be) may reveal a perturbation over the interval covered by a series of observations. If the interval exceeds the period of the perturbation (assuming for the moment one unseen companion only), the period P will be revealed by a repetition of the pattern of the remainders. There are several cases on record where the series of observations have covered more than two periods. If the remainders show that one full period has not yet been covered but if the amplitude of the observed portion of the perturbation is sufficiently large, an estimate of the period may be made and provisional results obtained for the other elements.

The dynamical elements period P, periastron passage T and eccentricity e may be derived from the time displacement curves R/t in any coordinate, ideally and preferably right ascension and declination, in the manner described in Section (e) of Chapter 10.

The method is of particular value for the photocentric orbit of an unresolved astrometric binary. The success of the method depends on the precision with which the proper motion of the barycenter is known, and can be allowed for. This generally implies coverage of the orbital motion over more than one period. If a substantial portion of the period has been covered and the observations are of sufficient precision, a provisional value of the proper motion of the barycenter may be obtained, allowed for and the method applied to obtain provisional values for the elements T and e.

Having plotted the remainders R after correcting for a proper motion of the barycenter, time-displacement curves are drawn that satisfy the remainders. Periastron and apastron are located by the fact that their epochs differ by half a period (or we may say their mean anomalies differ by 180°), and their ordinates are equal and of opposite signs when referred to the center of the orbit. Periastron and apastron are located by making a copy of a displacement curve, reversing it along the central line representing the center of the orbit, and shifting the reversed curve half a period along the time axis. Generally two pairs of intersections result: the single one on the shorter, steeper branch, and the middle intersection on the longer branch represent periastron and apastron respectively.

The slopes of the displacement curve represent projected velocities dR/dt. The ratio of the true (unprojected) velocity vectors at periastron and apastron is $-(1 + e)/(1 - e)$, their directions being opposite. This ratio does not alter in projection; hence the ratio of the slopes $(dR/dt)_P$ and $(dR/dt)_A$ at periastron and apastron respectively amounts to $-(1 + e)/(1 - e)$. We thus find the eccentricity of the orbit, independently of the barycenter, through the relation

$$e = \frac{(\mathrm{d}R/\mathrm{d}t)_P + (\mathrm{d}R/\mathrm{d}t)_A}{(\mathrm{d}R/\mathrm{d}t)_P - (\mathrm{d}R/\mathrm{d}t)_A}\,. \tag{13.1}$$

This method becomes unreliable when periastron and apastron are close to the extreme displacements, in which case a combination of graphical and analytical methods may be used (van de Kamp, 1947a).

The *geometric elements* (B), (A), (G), and (F) refer to the photocentric orbit of the unresolved system and are given in parentheses to distinguish them from the geometric elements B, A, G, and F (natural elements or Thiele–Innes constants) of the relative orbit of primary and companion (Chapter 10). The geometric elements of the photocentric orbit are related to the Thiele-Innes constants as follows:

$$\begin{aligned} (B) &= -\frac{\alpha}{a}B, \qquad (G) = -\frac{\alpha}{a}G, \\ (A) &= -\frac{\alpha}{a}A, \qquad (F) = -\frac{\alpha}{a}F, \end{aligned} \tag{13.2}$$

where α and a are the semimajor axes of the photocentric and relative orbit respectively, and assuming constant blending effect β throughout the entire series of observations.

The following formulas are used for the orbital displacements in the two equatorial coordinates:

$$\begin{aligned} &(B)x + (G)y \quad \text{in RA,} \\ &(A)x + (F)y \quad \text{in Decl.,} \end{aligned} \tag{13.3}$$

where x and y are the elliptical rectangular coordinates in unit orbit (Chapter 10).

The formulas for analyzing the path of primary or photocenter for proper motion, parallax and quadratic time effect, and orbital motion are therefore

$$X = c_X + \mu_X t + q_X t^2 + \pi P_\alpha + (B)x + (G)y$$

and (13.4)

$$Y = c_Y + \mu_Y t + q_Y t^2 + \pi P_\delta + (A)x + (F)y.$$

Generally the observed path is first analyzed for proper motion and parallax, and quadratic time effect if need be. Hence, after P, T, and e have been determined, they may be used via the elliptical rectangular coordinates in unit orbit, x and y, to calculate the geometric elements (B), (G), (A), and (F) and corrections to c, μ, q, and π, if desired. As in the case of relative orbits for visual binaries (Chapter 10), the conventional elements α, i, ω, and Ω may be derived from the natural elements through the following relations:

$$(A)^2 + (B)^2 + (F)^2 + (G)^2 = 2k \qquad (13.5)$$

$$(A)(G) - (B)(F) = m \qquad (13.6)$$

$$j = (k^2 - m^2)^{1/2} \qquad (13.7)$$

from which

$$\alpha = (j + k)^{1/2} \qquad (13.8)$$

$$i = \arccos m\alpha^{-2}. \qquad (13.9)$$

Here the geometric elements and α are expressed in seconds of arc. Also

$$\omega + \Omega = \arctan \frac{(B) - (F)}{(A) + (G)}, \qquad (13.10)$$

where $\sin(\omega + \Omega)$ has the same sign as $(B) - (F)$, and

$$\omega - \Omega = \arctan \frac{-(B) - (F)}{(A) - (G)}, \qquad (13.11)$$

where $\sin(\omega - \Omega)$ has the same sign as $-(B) - (F)$. The quadrants of $\omega + \Omega$ and $\omega - \Omega$ are unambiguously determined. However, we may add 360° to $\omega + \Omega$ (or to $\omega - \Omega$) and we thus obtain two solutions

$$\omega, \Omega \quad \text{and} \quad \omega \pm 180°, \quad \Omega \pm 180°.$$

Unless a distinction is possible, the solution is adopted corresponding to $\Omega < 180°$.

At this stage one may wish to make another determination of α and its error, using formulae (11.10), with orbital factors based on the obtained orientation elements Ω, ω, and i (formulae (11.11) and (11.12)).

(c) *Mass-function. Orbital constant. Dynamical interpretation.* We assume that the observational material of the unresolved binary has been adequate to separate the stars' own and its annual parallactic motion. Generally there will not have been any problem provided the period of the star's own orbit is not too close to one year. With a known value of the (absolute) parallax p'', the angular value α'' of the semi-axis major of the perturbation orbit is converted to the linear value in astronomical units:

$$\alpha_{(AU)} = \frac{\alpha''}{p''} \qquad (13.12)$$

The sum of the masses of the two components is given by the harmonic relation between mass, space and time:

$$\mathcal{M}_A + \mathcal{M}_B = \frac{a^3}{P^2}, \tag{12.1}$$

where $\mathcal{M}_A$ and $\mathcal{M}_B$ are the masses, a the linear value of the semi-axis major and P the period of the relative orbit of the two components, all expressed in *astronomical units*, i.e. solar mass, semi-axis major for the Earth orbit around the Sun and sidereal year. While the harmonic relation furnishes the total mass of a binary-system if a and P are known, such obviously is not the case for the unresolved binary which yields α instead of a. What information can we derive and how?

First let us consider the case where the component of mass $\mathcal{M}_B$ is truly invisible, i.e. contributes no light to the image of the central star. The values of $\mathcal{M}_A + \mathcal{M}_B$ and a are not known *a priori*; an astrometric analysis only yields values for the period P and for the semi-axis major α of the perturbation orbit of the visible star A around the center of mass of the two components, i.e.

$$\alpha = a_1 = \mathrm{B}a, \tag{11.3}$$

where B is the fractional mass $\mathcal{M}_B/(\mathcal{M}_A+\mathcal{M}_B)$ of the unseen companion in terms of the combined mass of primary and companion (Chapter 12). The two observed quantities α and P yield a value for the *mass function*

$$\frac{\alpha^3}{P^2} = \mathrm{B}^3(\mathcal{M}_A + \mathcal{M}_B) \tag{13.13}$$

which may also be written

$$\mathcal{M}_B = \alpha P^{-2/3}(\mathcal{M}_A + \mathcal{M}_B)^{2/3}. \tag{13.14}$$

Evidently the basic dynamical information furnished by the astrometric analysis is the quantity

$$\alpha P^{-2/3} \tag{13.15}$$

which we name *orbital constant*. Frequently the determination of α is less precise than that of the period P, particularly for a perturbation of small amplitude. But, as in the case of visual binaries, one may expect a reduction in the error of the orbital constant due to partial cancelling of errors in α and in P. The orbital constant together with an adopted value for $\mathcal{M}_A$, yields a value of $\mathcal{M}_B$.

Generally, however, we do not know whether the observed orbit refers to the pure image of the primary or to the photocenter of primary and companion. To take into account a possible blending effect we recall that

$$\alpha = (\mathrm{B} - \beta)a \quad \text{or} \quad \mathrm{B} = \frac{\alpha}{a} + \beta \tag{11.7}$$

and the following mass function is found:

$$\frac{\alpha^3}{P^2} = (\mathrm{B} - \beta)^3 (\mathcal{M}_A + \mathcal{M}_B) \tag{13.16}$$

which may also be written as

$$\mathcal{M}_B = \alpha P^{-2/3}(\mathcal{M}_A + \mathcal{M}_B)^{2/3} + \beta(\mathcal{M}_A + \mathcal{M}_B). \tag{13.17}$$

Again B is the fractional mass and β the fractional luminosity of the companion B in terms of the combined luminosity of A and B (Chapter 11). An adopted value of the combined mass yields a lower limit (13.17) for the mass of the companion.

To arrive at acceptable values for the masses, consider the relations (12.1) and (13.17). With known P one adopts a range of values for $\mathcal{M}_A + \mathcal{M}_B$, i.e. a, and for β, and one obtains a range in values of $\mathcal{M}_A$ and of $\mathcal{M}_B$. The different values of a will limit the possible range of $\mathcal{M}_A$, $\mathcal{M}_B$ and of β; a larger value of a and/or a small value of Δm (large β) might have resulted in earlier visual detection of the companion. The various constraints generally lead to rather narrow limits for the mass of the unseen companion and to an upper limit for its luminosity.

Of special interest is the case of an astrometric binary with an unseen companion of very small mass and low luminosity. This would be indicated by a very small value of the orbital constant $\alpha P^{-2/3}$ so that to a high degree of approximation (13.17) may be written as

$$\mathcal{M}_B = \alpha P^{-2/3}\, \mathcal{M}_A^{2/3} + \beta \mathcal{M}_A \tag{13.18}$$

or even

$$\mathcal{M}_B = \alpha P^{-2/3}\, \mathcal{M}_A^{2/3} \tag{13.19}$$

for near zero luminosity such as would be the case for a feeble substellar or a planetary companion. In the latter case (12.1) becomes

$$\mathcal{M}_A = \frac{a^3}{P^2}\,. \tag{13.20}$$

For an adopted value of $\mathcal{M}_A$ one finds $\mathcal{M}_B$ from (13.19). In this case the principal sources of error in $\mathcal{M}_B$ are the observational errors of α and of the adopted value of $\mathcal{M}_A$. An error of 10% in $\alpha P^{-2/3}$ results in an error of the same amount in $\mathcal{M}_B$. An uncertainty of 10% in $\mathcal{M}_A$ results in an error of 7% in the value of $\mathcal{M}_B$. For the same values of $\mathcal{M}_B$ and a, the linear value of the semi-axis major α of the perturbation is inversely proportional to the mass $\mathscr{M}_A$ of the visible component i.e. α is much larger for a primary of small mass, such as a red dwarf, than for a star of rather larger mass such as our Sun.

(d) *Once more: systematic errors.* The smaller the perturbation the more delicate the problem; while the attainable accuracy may be 0.″002, systematic 'breaks' may be larger. By now some twenty perturbations have been found photographically, with total amplitudes larger than 0.″05, above the level of suspected or extreme established systematic errors. It is important to have series of observations without conspicuous gaps; continuous observational coverage must be aimed for, either as an *a priori* policy or as soon as a perturbation is established by accidental discovery. And, as mentioned before (Chapter 8), any quadratic time effect should be eliminated, to remove any contamination of the perturbation.

The two stars, Ross 614 and VW Cephei, have perturbations well over 0.″1 and did not remain unresolved for very long; they are now resolved, *bonafide* 'visual' binaries. There are, however, several other cases of perturbations of a provisional, suspected, or uncertain nature. In most of these cases the total amplitude is small, hovering near or below the critical value of 0.″05 mentioned above. Hence the possible presence of systematic errors becomes a serious matter. One way to check the reality of a small perturbation is to confront it with the behavior of residuals derived from parallax (and proper-motion) determinations of a number of stars with no suspected perturbations. This may be done by averaging the residuals for several stars for different portions of the sky, such as different declination zones (Lippincott, 1971). Another approach is the thorough analysis of the path for any one star, using a variety of possible parameters (Hershey, 1973a). In any case, and above all, it is desirable to follow the observed paths of as many stars as possible, over as long an interval as possible, taking as many plates as possible. Obviously, at any one observatory this cannot be done for many stars. It is desirable that the paths of some of the same stars should be followed at several observatories.

As indicated in Chapter 3 a systematic equation may be recognized as a temporally limited deviation of non cosmic origin. It is hoped that the concern with systematic errors, a sign of developing maturity in our current studies, will be intensively pursued. Needless, if not trivial, to say, it is important to keep the optical performance of the telescope, as well as the observing techniques, as constant as possible.

(e) *Perturbations in visual binaries.* For several binaries the residuals from the best possible relative orbit have revealed systematic deviations, which point to a perturbation caused by a third component, close to one of the visible components. In these cases the systematic behavior of the residuals in both coordinates is analyzed for Keplerian motion, care being taken to provide corrections to the orbital elements of the 'large' orbit. After the orbital motion in the 'small' orbit has been determined, the original observations are corrected for the perturbation and the elements for the large orbit are recomputed. From the analysis it cannot be determined whether the third body is close to the primary or to the secondary component; this can be determined only by measuring the orbital motion on an astrometric background of other stars.

Although perturbations already had been noticed even in visual measures, it is obvious that the chances for discovery are increased by the photographic technique.

CHAPTER 14

UNSEEN ASTROMETRIC COMPANIONS. ILLUSTRATIONS

(a) *Review.* The classical discoveries of unseen companions from perturbations in visually observed paths are those of Sirius, Procyon, also of the visual binaries Zeta Cancri, Xi Ursae Majoris, and of Zeta Aquarii, partly based on photographic observations (Chapter 1) Over the past half century numerous perturbations have been discovered photographically both in stellar paths on a background of reference stars and from multiple exposures of relative positions of double star components. Early discoveries at the McCormick and the Allegheny observatories were followed by the intensive observing program on nearby stars begun in 1937 at the Sproul Observatory, which is contributing to the continuing discovery of unseen astrometric companions. Since well over a decade the United States Naval Observatory is now active in this field. Of the eight perturbations discovered at USNO, six have short periods, ranging from 7.2 to 0.87 yr.

The overwhelming majority of nearby stars are late-type dwarfs of low mass; the discovered unseen companions prove to be mostly objects of low mass, some sub-stellar, while two unseen companions appear to have masses less than that of Jupiter (Chapter 16).

As long as a companion remains unseen, only a lower limit and estimated range of its mass can be made. A rigorous mass and luminosity determination can be made only after visual detection yielding apparent separation and Δm of the components. Thus far two photographically discovered companions have subsequently been seen, namely Ross 614B and VW Cephei C. Together with the earlier visual detection of the companions of Sirius and Procyon, four perturbations have thus resulted in the discovery of resolved visual binaries and hence have yielded valuable information about stellar masses.

A summary of our knowledge of unseen astrometric companions was given by van de Kamp (1975), followed later by an updated survey, with an emphasis on possible direct detection by Lippincott (1978).

(b) *Illustrations.* From data accumulated up to the middle of 1980 we have selected a number of representative orbits, primarily from Sproul observations, which are briefly described and illustrated by time-displacement curves in RA and Decl. and by apparent orbits. For the Sproul series of plates

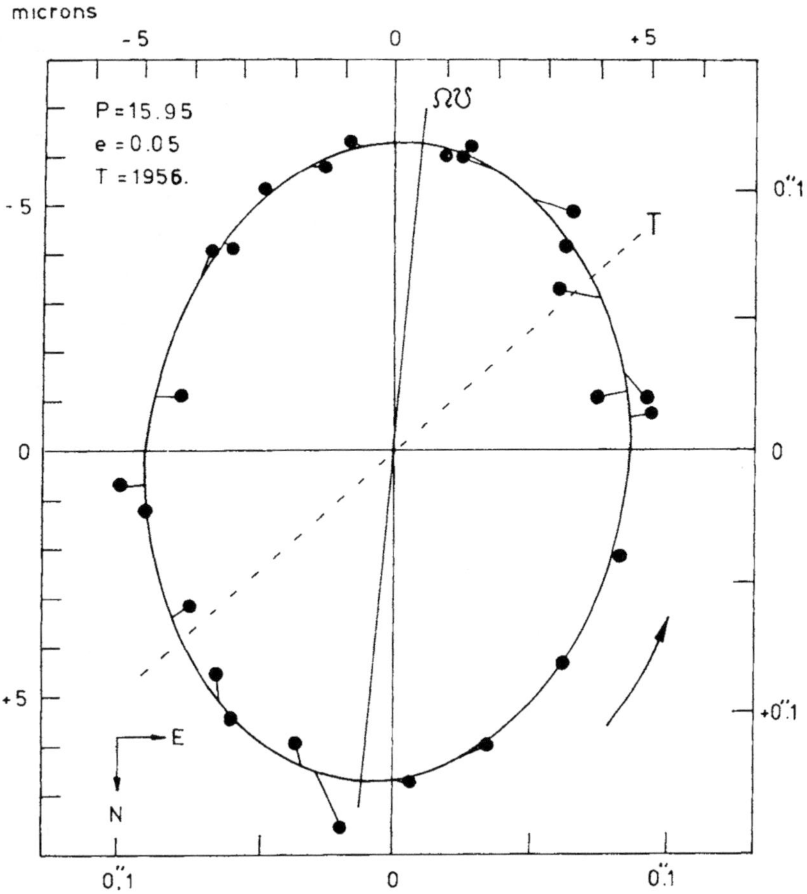

Fig. 14.1. BD + 66° 34 A. 0 29^{m}.3, + 66°58′, 10.5 dM, 2.5*e*, p = 0″.100. The perturbation was discovered by Alden (1947) as a by-product from an attempt to determine the mass-ratio of *A* and *B* at McCormick observatory.

Measurements of Sproul plates taken on 163 nights over the interval 1938–72 yield a period of P = 15.95 ± 0.22 yr, and α = 0″.125 ± 0″.002 for the photocentric orbit of BD + 66° 34 A around the barycenter of *A* and the unseen companion a (Hershey, 1973b). The orbit yields $\mathcal{M}_a = 0.13 \pm 0.02\mathcal{M}_\odot$, for the unseen companion assuming $\Delta m >$ 3.5, confirmed by Worley's failure to see the companion with the USNO 155-cm reflector. The unseen companion is probably a very late type M dwarf with an absolute visual magnitude > 14.

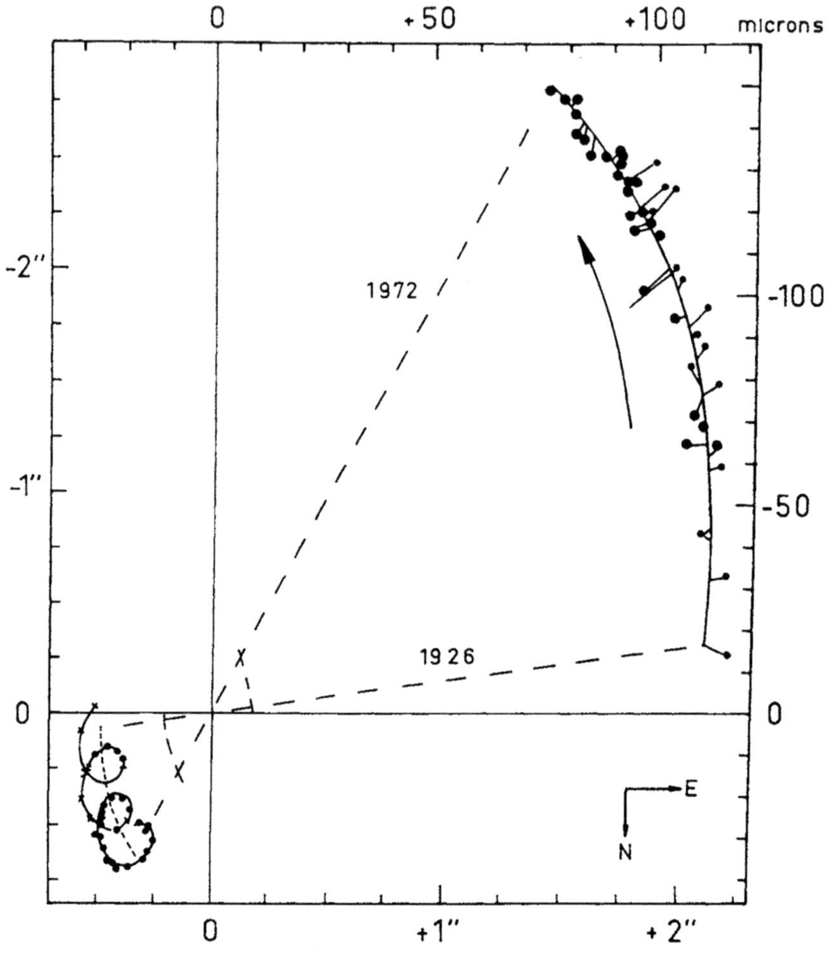

Fig. 14.2. BD + 66° 34 A, B. A provisional period of 320 yr has been determined for *Aa* and *B*. One of two possible cases of orientation makes the two orbits coplanar and corevolving. The mass-ratio determined for the 'long' orbit yields $\mathscr{M}_{Aa} = 0.53\ \mathscr{M}_\odot$, $\mathscr{M}_B = 0.12\ \mathscr{M}_\odot$. It illustrates the positions of BD + 66°34 (also named Mlb 377 and Vys 2) from 1926 to 1972. The origin of the plot is the computed center of mass of the *Aa*, *B* orbit. The arcs surrounding the center of mass point show the probable error of the center of mass. The large points are photographic positions, and the small points represent visual observations. Positions on the helix, where early visual measurements were made are shown by small crosses (×). The small dashed line in the helix, is the path of the center of mass of the *A*, *a* orbit.

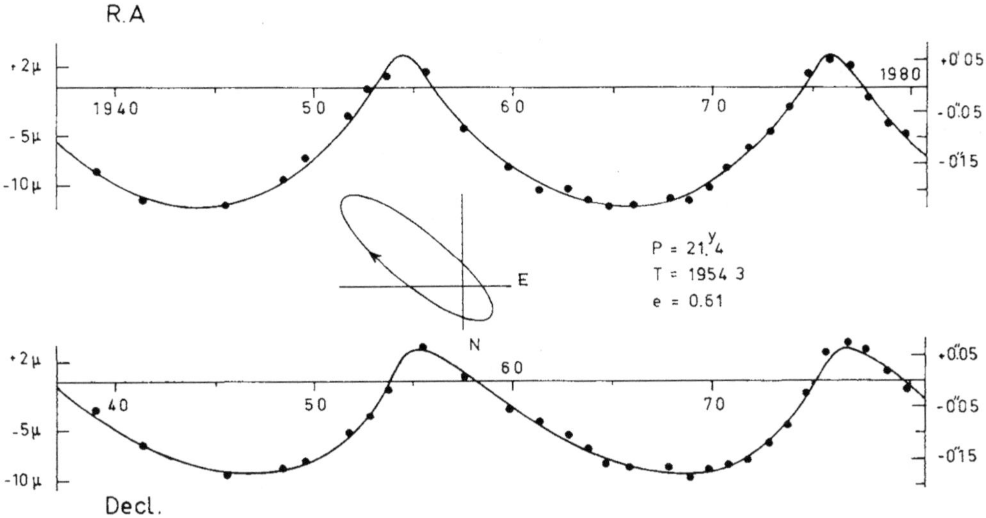

Fig. 14.3. *Mu Cassiopeiae.* $1^h4^m.9$, $+54°40'$, 5.2, G5VI, $p = 0''.130$. The perturbation was discovered by Wagman as a by-product of two parallax determinations at Allegheny Observatory. A recent study of Sproul plates taken on 215 nights over the interval 1937–1979 yields $P = 21.43 \pm 0.05$ yr, $\alpha = 0''.186 \pm 0''.001$, and a minimum mass for the unseen companion of 0.17 $\mathcal{M}_\odot$ with a $\Delta m > 5$ indicating a red dwarf (Lippincott, 1980).

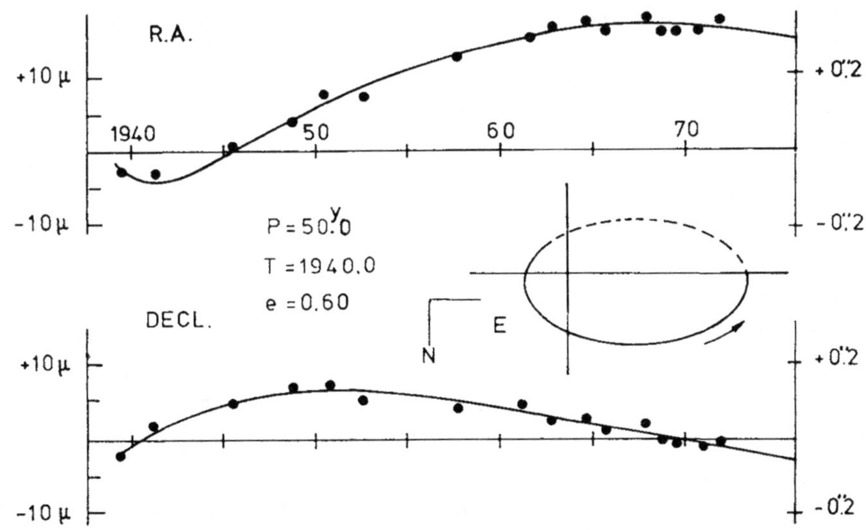

Fig. 14.4. BD $+6°398$, $2^h33^m.3$, $+6°39'$, 5.8, K3V, $p = 0''.128$. Discovered at Sproul Observatory in 1968, a study based on plates taken on 117 nights over the interval 1937–1973 yields provisional elements $P = 50$ yr, $\alpha = 0''.222 \pm 0''.008$; for $\Delta m > 5$ the mass of the unseen companion is 0.10 $\mathcal{M}_\odot$ (Lippincott, 1973).

A study at McCormick Observatory spanning 56 nights over the interval 1915–1974, yields $P = 60$ yr, $\alpha = 0''.257 \pm 0''.008$ and a mass of 0.12 $\mathcal{M}_\odot$ for the unseen companion (Martin and Ianna, 1975).

This is a good example of reliable provisional results obtained with two different telescopes, based on a sufficient portion of a not yet completed orbit.

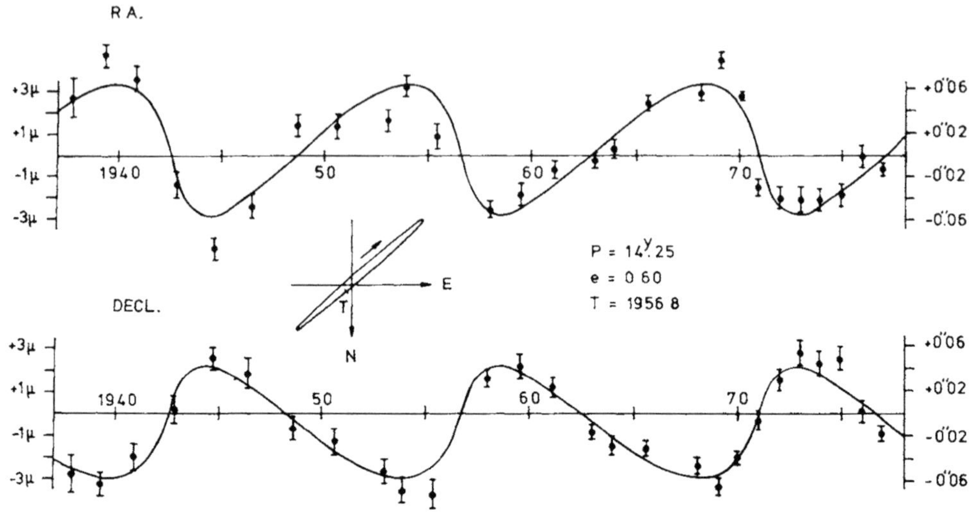

Fig. 14.5. χ^1 Orionis. $5^h51^m.4$, + 20°16′, 4.4, G0V, p = 0.″102. Discovered in 1977 at Sproul Observatory (Lippincott and Worth, 1978). Plates taken on 113 nights over the interval 1937 – 1977 yield P = 14.25 yr, α = 0.″095 ± 0.″003 and a minimum mass of 0.17 $\mathcal{M}_\odot$, for the unseen companion which very likely is a red dwarf. An example of a perturbation which hardly could have been found from observations covering less than a decade.

This is the second nearby naked-eye solar type star, recently found to have an low-mass unseen companion, the other one being PGC 372, $1^h38^m.7$, + 42°22′, 5.0, G2V, p = 0.″078 (Lippincott and Lanning, 1976).

The solar type star Zeta Cancri C (parallax 0.″042) has an unseen companion with a mass of 0.9 $\mathcal{M}_\odot$, apparently a white dwarf.

generally an average of three to four plates per night were taken; the scale value is 1 mm = 18.″87. The nightly mean remainders from a proper motion and parallax solution have been combined into normal points, over suitable time intervals so as not to obscure any 'short' period perturbations. The length of the vertical bars for the normal points in the Sproul displacement curves is twice the probable error.

Frequently annual normal points, extending over at most a six months' observing season, have been used. However final analyses for geometric elements were usually made from nightly remainders. While all observations should be continued with the telescope of discovery or started with other instruments, this is especially desirable for perturbations of small amplitudes.

The illustrations are arranged in order of RA; the coordinates are for the epoch 1950. Apparent visual magnitude, spectrum and parallax are given for each star.

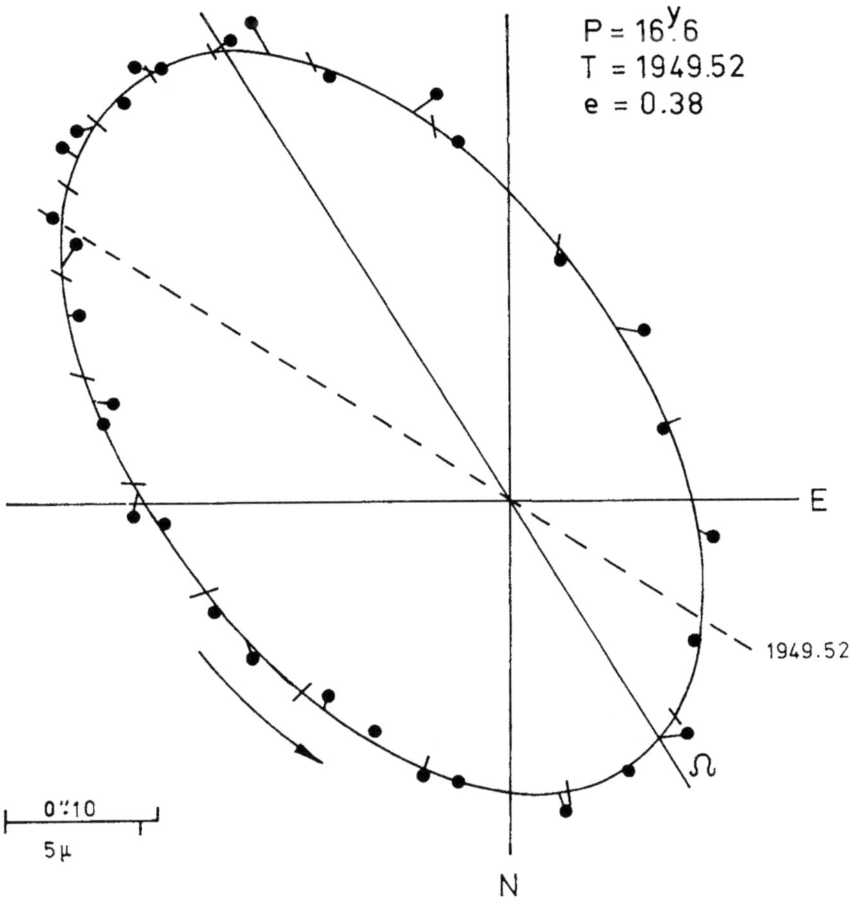

Fig. 14.6. Ross 614. Apparent photocentric orbit.

Figs. 14.6 and 14.7. Ross 614. $6^h26^m.8$, $-2°46'$, 11.3, M3e, $p = 0''.243$. Although now a *bona fide* resolved visual binary, this star is included because of historical interest. This was the earliest and up to now is the most accurate perturbation found by long-focus photographic astrometry. The classical discovery by Reuyl (1936) from only 25 McCormick plates over the interval 1927–1936, followed by later Sproul study (Lippincott, 1951), ultimately led to the visual detection and photographic record of the companion by Walter Baade in 1955 with the Hale 5-m reflector (Lippincott, 1955b).

Ross 614 has been the subject of several studies. The illustrations are from a 'definitive' study by Lippincott and Hershey (1972) based on Sproul plates taken on 252 nights over the interval 1938–1972 resulting in $P = 16.60$ yr, $\alpha = 0''.312 \pm 0''.002$.

A subsequent study by Probst (1977) is based on McCormick plates on 142 nights over the interval 1928–1975.

The few visual observations suggest $\Delta m = 3.5$, and a semi-axis major of $0''.91 \pm 0''.04$ for the relative orbit of *B* and *A*, leading to a mass of 0.062 $\mathcal{M}_\odot$ for the *B*-component, at present the lowest value for a well-determined mass of a visible star.

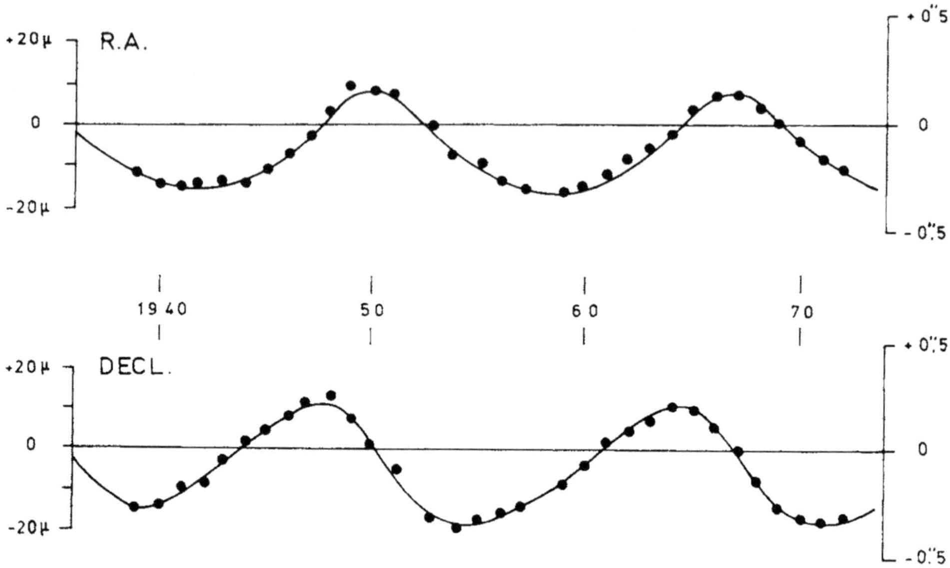

Fig. 14.7. Annual normal points and orbital displacement curves in RA and Decl.

The illustrations are followed by a number of selected descriptions, not illustrated, which include the three classical perturbations in the visual triple systems Zeta Cancri and Xi Ursae Majoris and in the visual binary Zeta Aquarii.

Stein 2051 A. $4^h26^m.8$, $+58°53'$, 11.1, M5, $p = 0''183$.
A combined astrometric study from Sproul and USNO plates by Strand reveals a perturbation in the A component of this nearby wide binary with current separation 7″. The period of 23 yr and semi-axis major $0''07$ are interpreted as due to a mass of $0.02\mathcal{M}_\odot$ for the unseen companion *a* (Strand, 1977).

G 107–69, 70. 7^h27^m1, $+48°13'$, 13.3, M5, 14.4, DC, $p = 0''090$.
This wide binary, now a quadruple system, under study at the USNO is of particular interest. The system has a parallax of $0''0902 \pm 0''0014$ (m.e.), based on 118 plates taken with the USNO 155 cm reflector. The primary G107–69, proves to be an unresolved astrometric binary with the shortest known period of 0.94 yr and a semi-axis major of $0''013 \pm 0''003$ (m.e.). For the secondary G107–70 which is partially resolved, $P = 20.5 \pm 1.9$ yr, $\alpha = 0''040 \pm 0''003$. The components of G107–69 are faint red dwarfs with masses $0.17\mathcal{M}_\odot$ and $0.08\mathcal{M}_\odot$. The components of G107–70 are white dwarfs (Harrington, Christy and Strand, 1981).

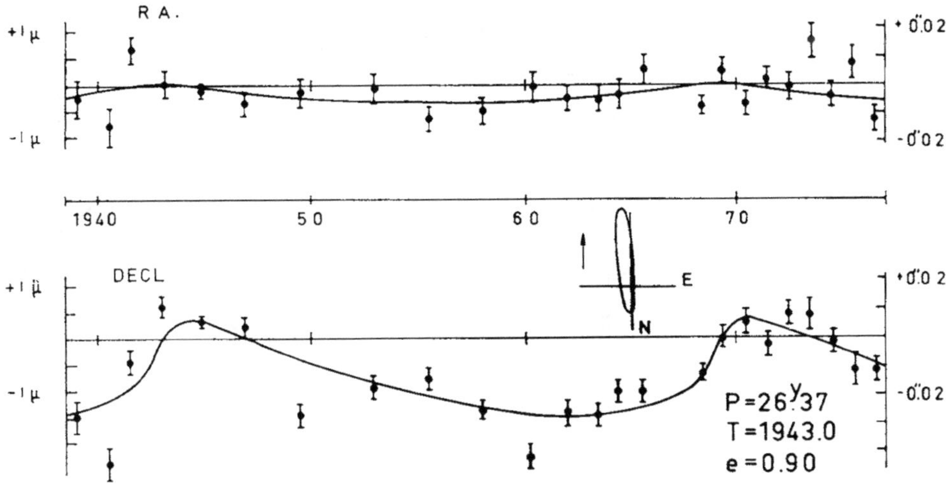

Fig. 14.8. BD + 68° 946, $17^h36^m.7$, + 68°23′, 9.1, M3.5V, p = 0″.213. Discovered at Sproul Observatory (van de Kamp and Lippincott, 1949; Lippinott, 1967), a recent study by Lippincott (1977) from plates taken on 287 nights over the interval 1937 to 1976 yields p = 26.37 yr. α = 0″.033 ± 0″.002 with a highly eccentric orbit (e = 0.9). The corresponding mass of the unseen companion most likely is not more than $0.01\mathcal{M}_\odot$.

Zeta Cancri C. $8^h9^m.3$, + 17°48′, 6.7, G2, p = 0″.042.
The visual binary Zeta Cancri AB has a period of 59.7 yr and a semi-axis major of 0″.88. The orbit of the distant companion Zeta Cancri C with respect to Zeta Cancri AB has a period of about 1150 yr and a semi-axis major 8″.0; Zeta Cancri C, a solar type star (G2) has a well established perturbation with P = 17.5 ± 0.2 yr, α = 0″.191 (Gasteyer, 1954).

This is the first classical discovery (1888) from micrometer observations of a perturbation in a visual binary (Seelinger, 1914). Assuming no light for the still unseen companion, its mass is found to be $0.9\mathcal{M}_\odot$. This fourth component of the by now quadruple system appears to be a white dwarf.

Xi Ursae Majoris A. $11^h15^m.5$, + 31°49′, 4.3, G0V, p = 0″.130.
The visual binary Xi Ursae Majoris AB has a period of 59.9 yr and a semi-axis major of 2″.54 (van den Bos, 1928), values not significantly changed by later orbit determinations. From micrometer observations of Xi UMa AB, Nörlund (1905) found this, the second classical discovery of a perturbation in a binary, which had been noted by Wright (1900) from spectroscopic observations. The perturbation was confirmed from multiple exposure photographs by Hertzsprung (1919): P = 1.8 yr, α = 0″.052. The unseen companion has an estimated mass of $0.3\mathcal{M}_\odot$. The fainter component Xi UMa B is a spectroscopic binary, thus making Xi UMa a quadruple system.

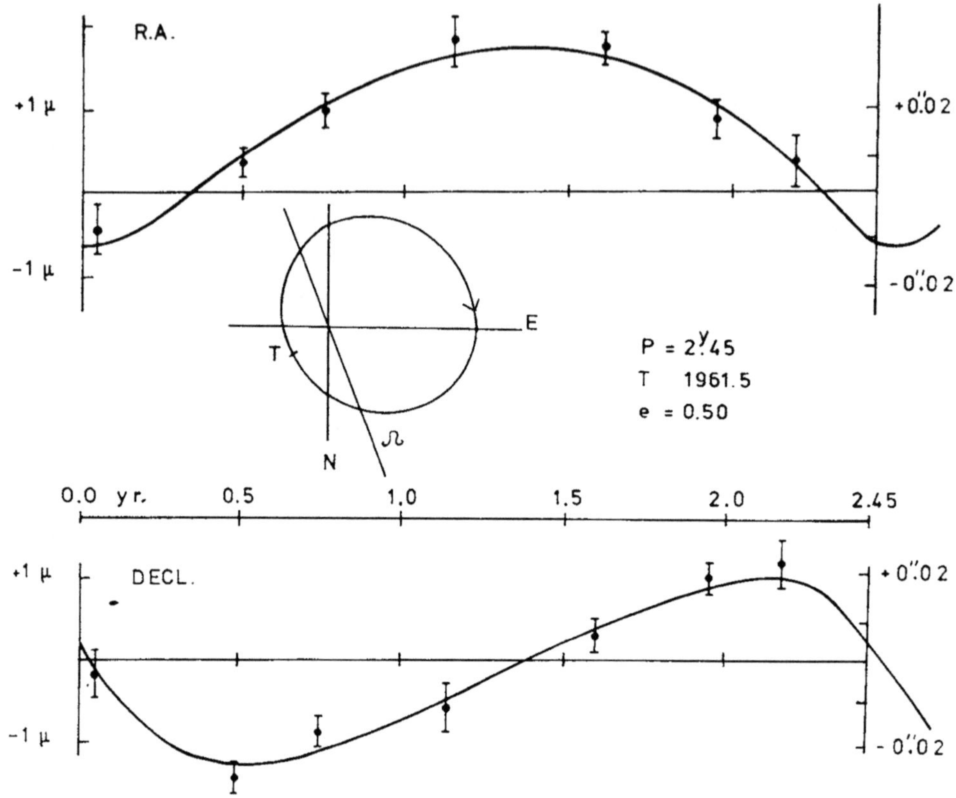

Fig. 14.9. Wolf 1062. $19^h9^m.6$, + 20°48′, 11.1, M4, p = 0.″120. Separate analyses from USNO plates over an interval of 5 years (1971–1976) by Harrington (1977) and over an interval of 38 years (1939 – 1976) from Sproul plates taken on 81 nights, (Lippincott, 1977) yield results in close agreement:

USNO P = 2.3 yr, α = 0.″031 ± 0.″002,

SPROUL P = 2.45 yr, α = 0.″026 ± 0.″002.

The mass of the unseen stellar companion has an estimated upper limit of 0.15 $\mathscr{M}_\odot$.

In the illustration based on the Sproul data the abscissae are counted in years from perihelion passage.

CC 986. $16^h22^m.7$, + 48°28′, 10.3, M3, p = 0.″136.

Discovered in 1977, an analysis from Sproul plates taken on 125 nights over the interval 1938 – 1977, yields P = 3.72 yr, α = 0.″049 ± 0.″002 and a range in minimum mass for the companion of 0.06 to 0.08$\mathscr{M}_\odot$ (Lippincott and Borgman, 1978).

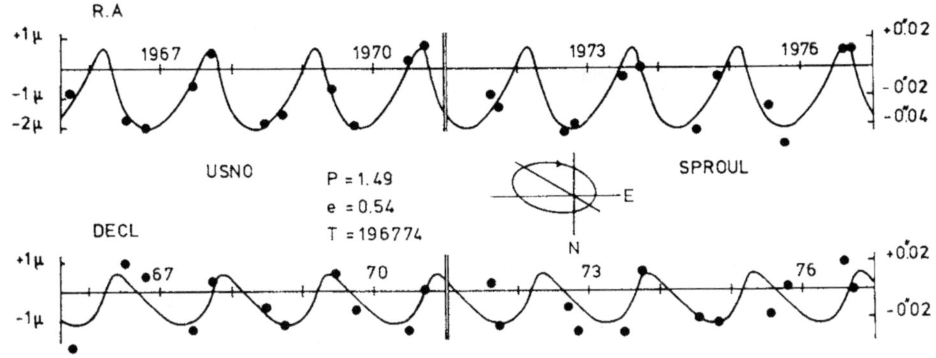

Fig. 14.10. G24–16, $20^h27^m.4$, $+9°31'$, 13.0, dM, $p = 0''.115$. The perturbation of this red dwarf, first noted by Riddle from plates taken with USNO 155 cm refractor, was analyzed by Harrington (1971) and yielded $P = 1.5$ yr, $\alpha = 0''.029$ from 88 plates taken over the interval 1965.7–1970.7. A subsequent study by Hershey (1977) from Sproul plates taken on 54 nights over the interval 1971–1976 yielded $P = 1.49$ yr, $\alpha = 0''.028 \pm 0''.004$, in good agreement with the USNO result. The estimated mass of the companion ranges from 0.07 to 0.11 $\mathscr{M}_\odot$.

The message from these studies is that a short-period perturbation for a nearby star can be discovered and analyzed in a time span of only a few years, assuming intensive and accurate observational coverage.

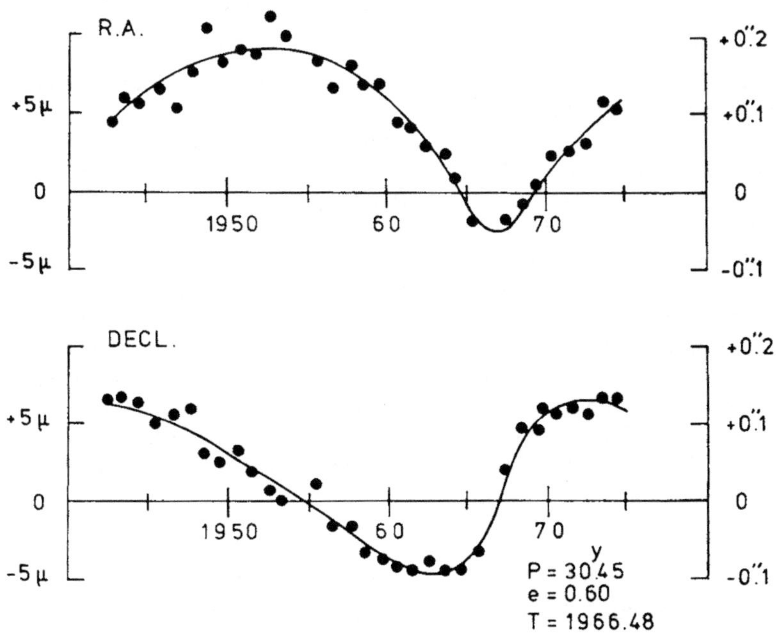

Fig. 14.11. VW Cephei. Yearly or semi-yearly normal points and calculated orbital displacement relative to the center of mass of VW Cephei AB–C.

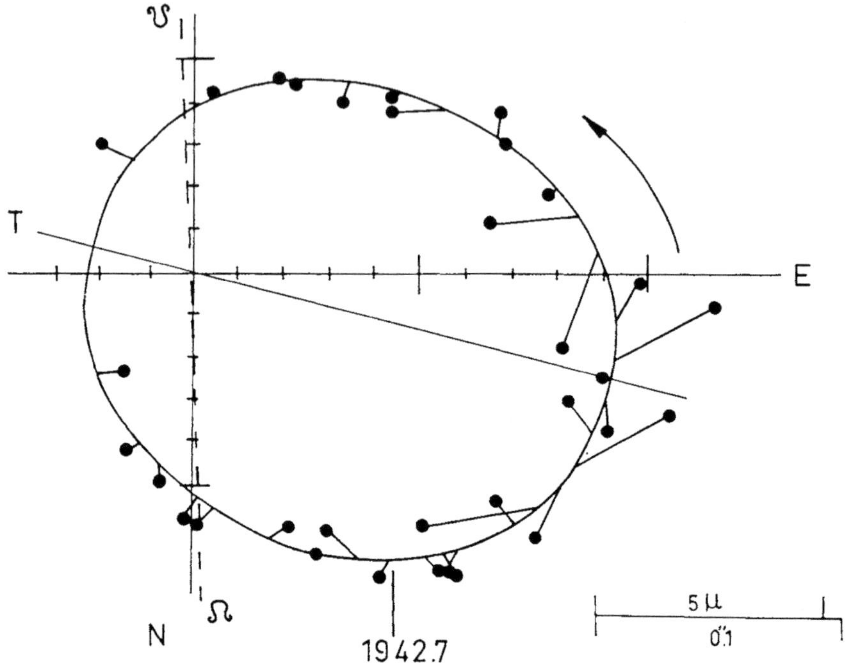

Fig. 14.12. VW Cephei. Apparent photocentric orbit.

Figs. 14.11 and 14.12. VW Cephei. $20^h38^m\!.0$, $+75°25'$, 7.3, G5, $p = 0.''041$. As in the case of Ross 614, although this is now a resolved visual binary, this star is also presented because of its unique historical interest.

VW Cephei is a W UMa type close eclipsing variable with a period of 0.278 days. An astrometric analysis by Hershey (1975a) from 610 Sproul plates taken on 164 nights over the interval 1952–1973 yields $P = 30.4$ yr, $\alpha \pm 0.''130 \pm 0.''002$; the inferred companion C was subsequently seen close to the predicted location at an observed separation $0.''64$ and an estimated $\Delta m = 2.9$. The resulting mass for the now visible companion is $0.58 \pm 0.14\,\mathcal{M}_\odot$. Ironically the behaviour of light-time residuals, (Payne-Gaposchkin, 1941), which prompted the long-term astrometric surveillance has turned out to be unrelated to a third body, being caused by a combination of both sudden and gradual changes in the period of AB. This star at a distance over 20 parsec was put on the observing program for the wrong reason. But the astrometric efforts nevertheless resulted in the discovery of a bona-fide companion for VW ,Cephei, at a distance over 20 parsec, well beyond the limit for any 'planned' discoveries of unseen companions to nearby stars. How many more similar situations do exist?

CC 1228. $20^h43^m\!.3$, $+44°19'$, 10.8, dM3, $p = 0.''082$.

Discovered at Sproul Observatory (Lippincott, 1979), measurements of plates taken on 79 nights over the interval 1939 to 1978, yield $P = 6.3$ yr, $\alpha = 0.''012 \pm 0.''002$. The most likely value for the mass of the unseen companion is $0.02\mathcal{M}_\odot$.

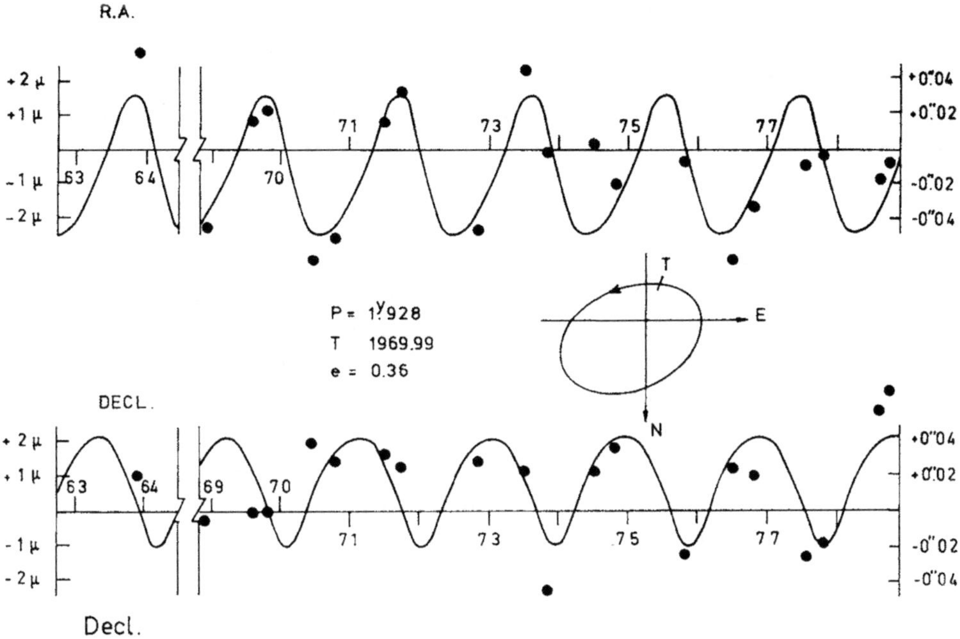

Fig. 14.13. Wolf 922. $21^h28^m\!.6, -10°1'$, 11.9, M4.5e, $p = 0''\!.122$. An analysis of Sproul plates taken on 53 nights over the interval 1963–1978 yields $P = 1.93$ yr, $\alpha = 0''\!.042 \pm 0''\!.003$. The minimum mass for the companion is $0.11\,\mathscr{M}_\odot$, possibly a faint red dwarf subluminous for its mass (Lippincott, 1979).

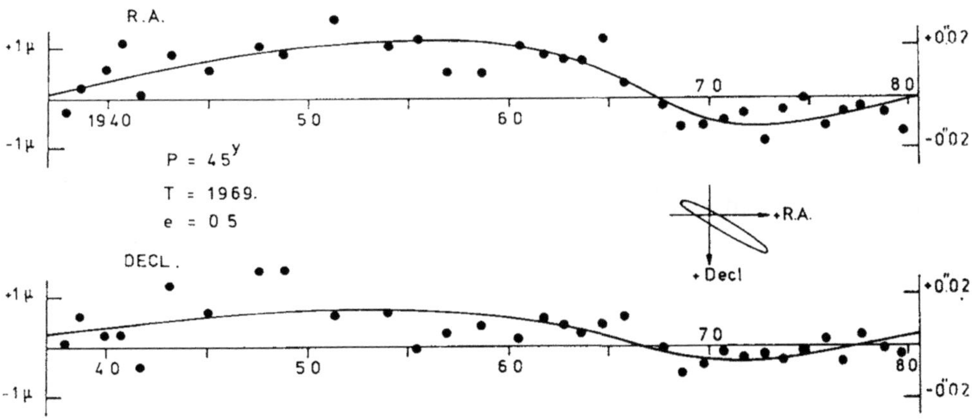

Fig. 14.14. BD + 43°4305, $22^h44^m\!.7$, +44°05′, 10.2, M4.5e, $p = 0''\!.200$. The perturbation of this star, also named EV Lacertae, was discovered at Sproul Observatory and analyzed in 1972 by van de Kamp and Worth. A recent updated study by van de Kamp and Lippincott (1980) based on plates taken on 322 nights over interval 1937 to 1979 yields $P = 45$ yr, $\alpha = 0''\!.0206 \pm 0''\!.001$, leading to a likely mass in the range 0.003 to $0.005\,\mathscr{M}_\odot$ for the unseen companion.

61 Cygni, $21^h04^m\!.7$, $+38°30'$, 5.2, K5e, 6.0, K7, $p = 0''\!.292$.
According to Deutsch and Orlova (1977), relative photographic positions of the two components from observations at Pulkovo and Sproul observatories indicate the probable presence of three unseen substellar companions, two of which are confirmed through a subsequent study by Deutsch (1978), taking also into account USNO data.

BD + 27°4120, $21^h35^m\!.8$, $+27°30'$, 9.8, M0e, $p = 0''\!.081$.
Discovered by Bieger (1964) from Sproul plates taken over the interval 1939–1963. A recent analysis by Lippincott and Turner (1980) from observations covering 179 nights from 1939 to 1979 may be represented by $P = 85$ yr, $\alpha = 0''\!.436 \pm 0''\!.004$. A mass range from 0.23 to $0.29\mathcal{M}_\odot$ is found for the unseen companion, suggesting a somewhat underluminous M dwarf.

The elements are provisional – the period may well be off by ten years. This is a good example of results requiring continued observations over an appreciable part of a century.

Zeta Aquarii B. $22^h26^m\!.3$, $-0°17'$, 4.6, dF1s, $p = 0''\!.043$.
This long-period binary *AB* was found to have a perturbation (Strand, 1942); the unseen companion belongs to the *B*-component (Franz, 1958). A study by Harrington (1968) yields

$$P_2 = 856 \text{ yr}, \; a = 5''\!.055 \text{ for the } AB \text{ orbit,}$$
$$P_1 = 25.5 \text{ yr}, \; \alpha = 0''\!.072 \text{ for the perturbation.}$$

The mass of the unseen companion is found to be $0.22\mathcal{M}_\odot$.

An astrometric study of the nearest star Alpha Centauri C (Proxima Centauri): $14^h26^m\!.3$, $-62°28'$, 11.0, M5e, $p = 0''\!.772$ by Kamper and Wesselink (1978) based on plates taken at the Yale (Johannesburg) and Cape observatories over the interval 1926 – 1976 reveals no measurable perturbation.

CHAPTER 15

UNSEEN ASTROMETRIC COMPANIONS. GENERAL

(a) *Mass-luminosity relation.* The masses of unseen astrometric companions generally are fairly well known within narrow limits, as are the upper limits of their luminosities. Figure 15.1 shows a mass-luminosity diagram from information kindly prepared by S. L. Lippincott.

The locations for M-dwarf components of known visual binaries have been

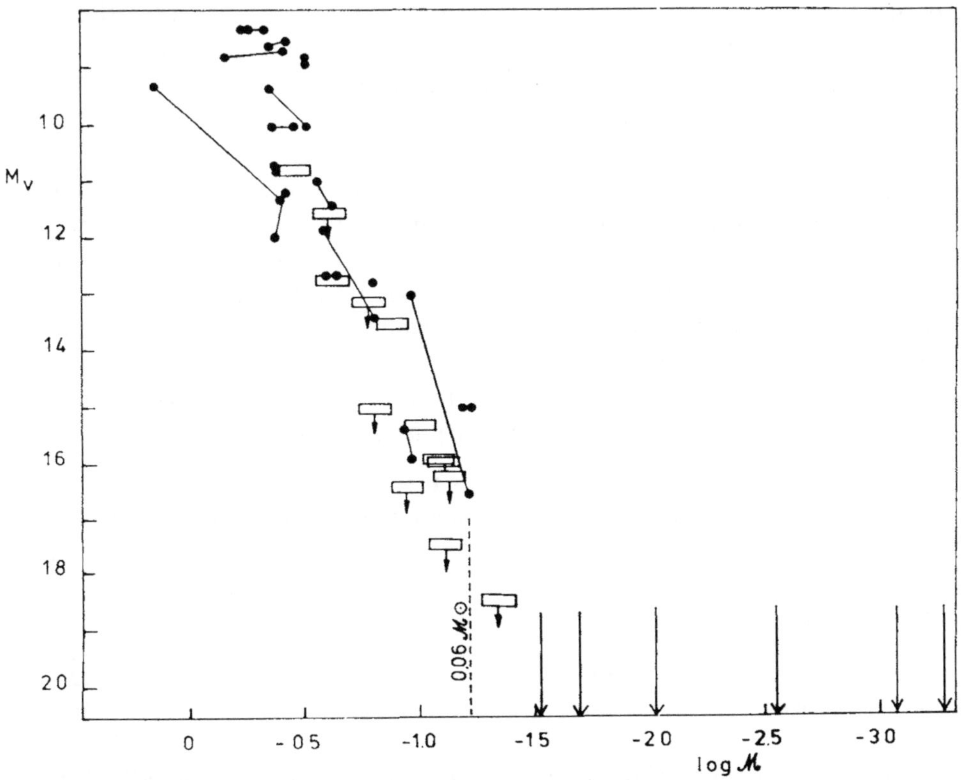

Fig. 15.1. Likely locations of unseen astrometric companions in mass-luminosity diagram. Known M-dwarf binary components are entered as dots, the unseen companions as rectangular boxes, whose locations in several cases are uncertain as to absolute magnitude. Complete lack of knowledge of absolute magnitudes exists for the objects below log mass $= -1.5$.

entered as dots. These include the formerly unseen companions Ross 614 B and VW Cep C, now visible stars located on the main sequence.

The unseen astrometric companions cannot be placed rigorously until their absolute magnitudes become known. Guided by an extrapolation of the known M-dwarf binary components, represented by $\log \mathcal{M} = 0.56 - 0.10\, M_v$, (Lippincott and Turner, 1980), the unseen companions have been entered as rectangular symbols for an appropriate and likely magnitude, compatible with the astrometric results. Several locations refer to highest adopted brightness as indicated by the attached arrows.

For a number of objects (CC 1228 B, Stein 2051 a, BD + 68° 946 B, BD + 43° 4305 B and the companions of Barnard's star) the absolute magnitudes are entirely undetermined; their locations are indicated by the downward pointed arrows in the lower right part of the diagram. Whether these very faint companions lie on an extension of the main sequence toward very low luminosities and very low masses remains an unsolved question at the moment.

(b) *Number- and mass-density*. An earlier analysis on this subject (van de Kamp, 1976) resulted in a provisional estimate of 0.21 ± 0.06 (p.e.) pc^{-3} for the number-density and $0.063 \pm 0.018 \mathcal{M}_{\odot}\ pc^{-3}$ for the mass-density in our stellar neighborhood for the then known astrometric companions of stars. Since then several more unseen companions have been found; a new analysis has been made and is here presented.

Twenty-eight unseen astrometric companions are known, with parallaxes ranging from 0″.547 to 0″.031. A summary is given in Table 15.1.

TABLE 15.1
Number N of unseen companions up to increasing distance limits r

Parallax p	Distance r	Total number N	Relative volume
>0″.17	< 6 pc	6	1
>0.10	<10	15	5
>0.06	<17	24	24
>0.03	<33	28	166

The incompleteness of the small sample is indicated by the rapidly decreasing total number of discoveries with increasing distance r. Completeness requires N to be approximately proportional to r^3; instead the observations show an approximate proportionality with r, suggesting an incompleteness factor of r^{-2}. Additional incompleteness is caused by the virtual lack of discoveries in the southern equatorial hemisphere.

The discovery chances of unseen companions depend on (1) the amplitude of the perturbation which, *ceteris paribus*, is proportional to the parallax, (2) the observational coverage which likely is greater for nearby stars. Both (1) and (2) are not inconsistent with the above incompleteness factor r^{-2}. Is it possible to ascertain at what distance incompleteness becomes effective? We shall again try, in a most provisional, tentative and formal manner, to establish the distance up to which the survey may be considered complete, and thus estimate the number-density up to that distance.

Constant space or number density is represented by

$$\log N = \log \tfrac{4}{3}\,\pi n + 3 \log r$$

or

$$\log N = 0.622 + \log n + 3 \log r. \tag{15.1}$$

Here N is the total number up to and including distance r; n is the constant number density per cubic parsec. The constant slope of this relation is

$$\frac{\mathrm{d} \log N}{\mathrm{d} \log r} = 3. \tag{15.2}$$

The observed slope for the present material is much less than 3, decreases with increasing distance and becomes horizontal at the distance beyond which no unseen companions have (yet) been found. For the current material this limit is $\log r = 1.52$, or $r = 33$ parsec; the two most distant unseen companions have observed parallaxes of $0\overset{''}{.}031$ (32 parsec).

As before, in order to arrive at an estimate of the number density n, the observed values of r and N are represented by

$$\log N = a + b \log r + c\,(\log r)^2 \tag{15.3}$$

with slope

$$\frac{\mathrm{d} \log N}{\mathrm{d} \log r} = b + 2c \log r \tag{15.4}$$

and the distance r_3, corresponding to complete coverage, would be obtained for slope 3, i.e.

$$\log r_3 = \frac{3 - b}{2c}. \tag{15.5}$$

A least squares solution from 28 equations yields

$$\log N = -\,0.79 + \underset{\pm\,0.12}{2.72} \log r - \underset{\pm\,0.16}{0.80} (\log r)^2 \tag{15.6}$$

whence

$$\log r_3 = -0.175 \pm 0.013, \qquad r_3 = 0.67 \pm 0.19 \text{ pc}$$

compared with the 1976 value of 0.29 ± 0.10 pc.

The values of a, b, and c yield $\log N_3$, and (15.1) leads to

$$\log n = a - 0.622 + (b - 3) \log r + c(\log r)^2 \tag{15.7}$$

whence

$$\log n = -1.39 \pm 0.06 \quad \text{or} \quad n = 0.041 \pm 0.006$$

compared with the 1976 value of 0.14 ± 0.04 for n. The present results for r_3 and n differ appreciably from the earlier results; the percentage errors have been reduced.

In view of the limited material, and r_3 appearing as an extrapolated value of the ad hoc analytical representation, these differences, nor the present results should be taken seriously. The results are provisional and selective; the analytical representation is nothing but a good try, and may serve to stimulate further work. As found before, observational incompleteness formally sets in before the distance to the nearest star.

To allow for the absence of material below −10° decl. we multiply the above value of n by 1.6, resulting in 0.070 ± 0.010 for the number density of unseen astrometric companions in our neighborhood. And, if we adopt an average value of $0.3\mathcal{M}_\odot$ for their masses, we arrive at a mass density of $0.021 \pm 0.003\mathcal{M}_\odot$ pc^{-3} (as compared with the earlier result 0.063 ± 0.018 $\mathcal{M}_\odot$ pc^{-3}).

The present result may be compared with the mass of visible stars in our neighborhood of $0.06\mathcal{M}_\odot$ pc^{-3} (Gliese, 1956; Luyten, 1968) and the total mass density of $0.14\mathcal{M}_\odot$ pc^{-3} derived from dynamical considerations by Oort (1965), considered as uncertain by the author of that study (1976).

Very provisionally we conclude that unseen companions form a minor but measurable fraction of the total mass-density in our galactic neighborhood.

CHAPTER 16

PLANETARY COMPANIONS, BARNARD'S STAR

(a) *Introduction.* A conventional observational distinction between stars and planets is the upper limit of $0.001\mathcal{M}_\odot$ for the most massive known planet, Jupiter, in our solar system and the lower limit $0.06\mathcal{M}_\odot$ for the mass of a *bona fide* star, e.g. Ross 614 B. While there are indications for the existence of objects of intermediate mass (Chapter 14), the potential discovery of as yet unseen planets orbiting stars other than the Sun remains of prime interest. At present it is not possible to 'see' a planet as large as Jupiter and shining by reflected light, even for the nearest star Alpha Centauri, though new techniques, beyond the conventional visual, may yield results in the not too distant future.

Meanwhile the perturbation approach is available and promising, especially for the discovery of a Jupiter like planet orbiting a nearby star of low mass i.e. a red dwarf. A planet with Jupiter's mass orbiting the solar type star Alpha Centauri A (parallax $0''.76$), with a period of twelve years and semi-axis major of 5 astronomical units would cause a perturbation with a range of $0''.0076$. The same planet orbiting Barnard's star (parallax $0''.55$) in the same size oribit, would cause a much larger perturbation, with a range of $0''.04$, since Barnard's star has only 1/7 times the mass of Alpha Centauri A.

In an earlier paper 'Planetary Companions of Stars' (1956) we pointed out that unseen companions with 10 times Jupiter's mass, at distances between 5 and 10 parsec, when attached to lightweight primaries with a mass of 0.2 $\mathcal{M}_\odot$, may be discovered by long-focus photographic astrometry, if they have periods of the order of half a century or more. For a limited number of nearer stars, within 5 parsec, such companions with shorter periods of the order of a decade could be discovered. Of particular interest is the nearest star in the northern hemisphere, Barnard's star, which has been studied intensively, has yielded interesting cosmic results, and also valuable information about long-range consistency of the observations. We shall now present an update on Barnard's star.

(b) *Barnard's star: History, General Data.* In June 1916, a 'small star with large proper motion' ($10''.31$ annually) and still the star of largest known proper motion, was discovered by Edward Emerson Barnard (1857–1923), distinguished amateur astronomer who at that time was a member of the staff of the Yerkes Observatory (Barnard, 1916a). Barnard found and

Fig. 16.1. *Proper motion of Barnard's star.* In this combined print of two photographs taken 11 months apart with the Sproul refractor, the later plate was shifted slightly (to the left) with respect to the first. The large motion of Barnard's star – 10.3 sec of arc per year – can actually be detected from photographs taken a few nights apart.

measured the proper motion from two plates taken with the 6-inch Willard lens of the Lick Observatory, 1894 August 24, and with the 6-inch Bruce lens of the Yerkes Observatory, 1916 May 30. The star was afterwards located on plates taken in 1904 and 1907 (Barnard, 1916b) and on Harvard plates as far back as 1888 and 1890 (Pickering, 1916). An abstract of the discovery was presented at the nineteenth meeting of the American Astronomical Society, held at Sproul Observatory, Swarthmore College 1916 August 30 – September 2 (Barnard, 1916c).

Immediately after the discovery of this 'runaway star', the observatories concerned with parallax observations, took series of plates and established Barnard's star to be the second nearest known stellar object, next to the Alpha Centauri system (and of course the Sun). It is the nearest known star in the northern equatorial hemisphere. By now the distance is known with considerable accuracy; we adopt $0''.547 \pm 0''.003$ (p.e.) for the absolute parallax. The tangential velocity of Barnard's star is 89 km s^{-1}, the radial velocity as deduced from Doppler shift in the spectrum, is -108 ± 2.5 km s^{-1}. The distance to Barnard's star, decreasing at the rate of 0.036 light years per century, will gradually diminish from its present value of 6.0 light years to a minimum of 3.85 light years about A.D. 11 800 (Figure 16.2), at which time the annual proper motion will be 25″. The apparent visual magnitude 9.5 is presently decreasing at the rate of 0.013 mag. per century, and will reach a maximum brightness of 8.5 at closest approach (Chapter 8).

Barnard's star was given high priority on the Sproul astrometric program. Intensive observations were begun in 1938 and are being continued at the annual rate of some thirty nights or more. Barnard's star proves to be a good test for attainable accuracy from intensive and extensive observational material. A determination of secular acceleration benefits immensely from increasing time interval; as a matter of fact, *ceteris paribus*, the weight of such a determination increases with the fifth power of the time, provided constant coverage is maintained (Chapter 8). The detection of possible perturbations, particularly if these prove to be composite, obviously requires intensive observations over as extended a time interval as possible.

The distribution in time of the Sproul material over the interval 1916–79 is as follows:

Interval	Total number of nights
1916–1919	21
38–41	118
42–48	210
49–57	185
58–66	211
67–79	420
Total	1165

The last five divisions correspond to the aforementioned instrumental adjustments (Chapter 3). Beginning 1938, as a rule four plates with up to five exposures each were taken on any one night. A large number of reference stars for Barnard's star is not available on the Sproul 13 × 18 cm plates due to the long exposure time originally required. The basic exposure time was 15 min in 1916, but is now less than 30 s. Longer exposure times could very well bring out up to 10 or even 20 reference stars; however, this would require a magnitude compensation, i.e., a reduction of the brightness of Barnard's star, a procedure and nuisance which thus far we have preferred to avoid. Also, the choice of reference stars is restricted by the very large proper motion of Barnard's star, 10″.31 or 0.54 mm yearly; over the interval 1916–80 Barnard's star has moved as much as 35 mm, in the focal plane with respect to the background stars. After careful consideration four reference stars were selected, and labeled 1, 2, 4, and 5 (star 3 is the unresolved binary Kuiper 84, and is not suitable as a reference star). As may be seen from Figure 16.3 by 1988 Barnard's star will have moved 'outside' the 124 reference system, i.e., the dependence or relative positional weight of star 2 will become negative. That day of reckoning can be postponed by introducing star 5, which appears on the Sproul plates since 1945, and has been regularly measured since then. The dependence of star 5 became positive in 1955 and crossed the value of +0.10 in 1972. The four-star background 1245 currently appears ideal; it reaches highest positional accuracy in 1989 and remains useful until well into the twenty-first century (Figure 16.4). However since star 5 does not appear on the plates till about 1945, and up to the present time there is still little differ-

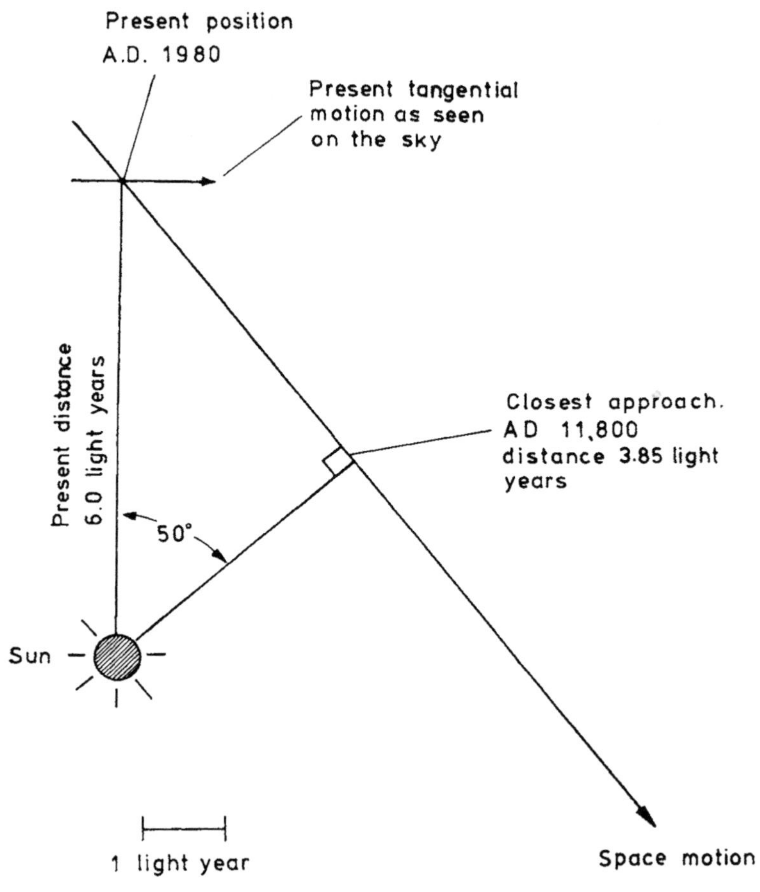

Fig. 16.2. Barnard's star. Path in space.

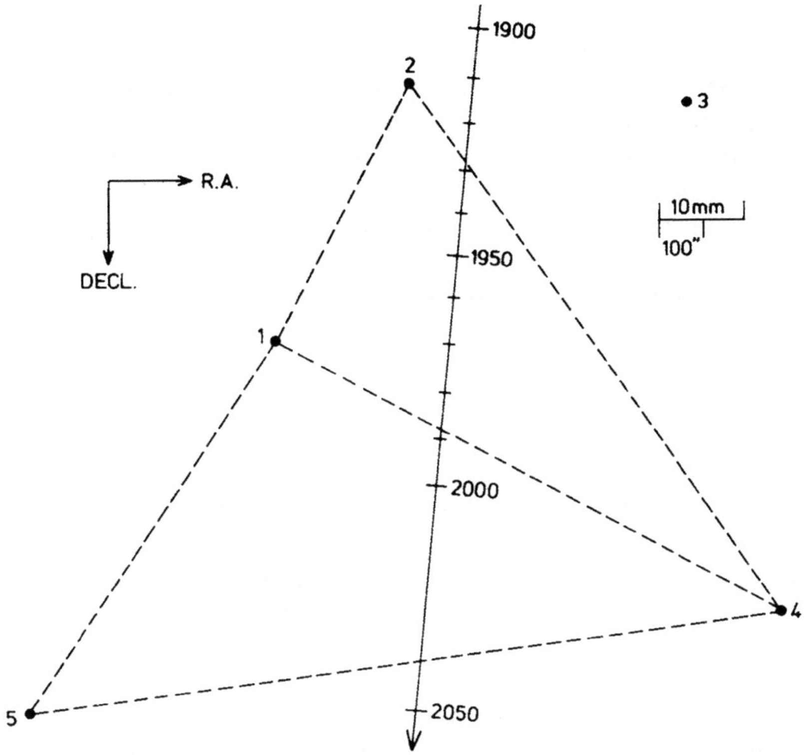

Fig. 16.3. Barnard's star. Path on sky with respect to reference stars.

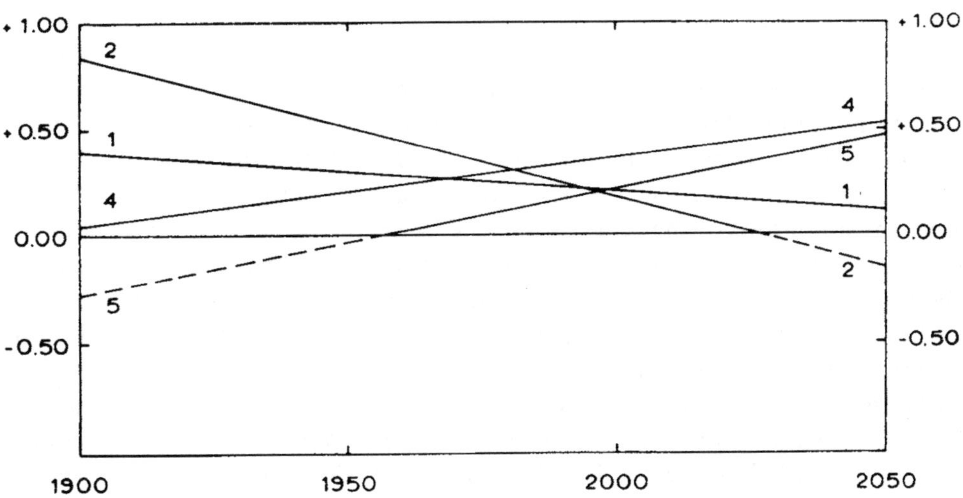

Fig. 16.4. Barnard's star. Dependence 'paths' of reference stars in the four-star configuration 1245.

ence between the 124 and 1245 background we have decided, at present, to use the 124 configuration (Figure 16.5).

For further details see Principles of Astrometry, Chapter 9, Section 6 (van de Kamp, 1967). All plates have been measured on the two-coordinate Grant measuring machine. Without let-up the intensive observational coverage of Barnard's star is being continued at Sproul Observatory. At the same time, as is wont to be, our critical sense grows faster than the accumulated observations. In particular, we have become aware of the potential effect of long-range systematic errors in our observations which are sharpened by the more accurate measurements.

(c) *Early results for perturbation.* After the initial parallax series of 1916–19 consisting of 24 plates, no plates of Barnard's star were taken at the Sproul Observatory until intensive coverage was begun in 1938. An early provisional, premature result was a short period spurious perturbation (van de Kamp, 1944), evidently suggested by the time-of-night error in declination, which is equivalent to a seasonal error since all observations are made close to the meridian. This effect amounts to about 0.4μ or 0ʺ008 around the middle of the night relative to dawn and dusk observations (Chapter 3). By 1956 the Sproul material showed evidence of a perturbation with an amplitude of several microns and a period of something like two decades. In 1963 we announced a perturbation with a semi-axis major of 1.3μ = 0ʺ024, a period of 24 yr, and an eccentricity of 0.6 (van de Kamp, 1963). By 1968, having added more observational material, the orbital elements were changed somewhat; semi-axis major 1.46μ, period 25 yr, and orbital eccentricity 0.75 (van de Kamp, 1969a). At that time I realized that the data could be equally well represented by two circular orbits, with periods of 26 and 12 yr, and with semi-amplitudes of 0.97μ and 0.39μ (van de Kamp, 1969b). The resulting companion masses were found to be 1.1 and 0.8 times the mass of Jupiter. Note for the first time the emergence of the shorter period. We realized later that the amplitude of the 'long' orbit, had been enhanced by the instrumental equation over the interval 1942–1948 (van de Kamp, 1975). Since then we take the point of view that the 1942–48 material in RA requires a correction of $+2.8\mu$, to bring about continuity with the earlier and later material. No appreciable correction in Decl. is indicated or required (Chapter 3).

Attempts to test the earlier Sproul results from 603 exposures obtained with the Allegheny and the Van Vleck (A,VV) refractors were unsuccessful (Gatewood, 1972; Gatewood and Eichhorn, 1973). The A,VV material is scattered unevenly over the interval 1916–71; compared with Sproul, the 34 A,VV normal points contain less than 5% as many exposures. The probable error of unit weight for the A,VV material is 0ʺ014, that of the Sproul material is 0″023. The difference may be partly due to the longer focal length of the Allegheny refractor, partly to the smaller number of exposures on the A,VV plates, so that the influence of night errors is possibly diminished; it is very doubtful that the large number of reference stars has played a role. Though the A,VV material appears to confirm the short period perturbation, at least in right ascension (van de Kamp, 1975), it is not sufficient either to confirm or refute the latest Sproul results.

The McCormick material, referred to in Chapter 8, is in satisfactory systematic agreement with the Sproul material; it is, however, too limited to test the Sproul perturbation results.

(d) *Latest Sproul solution for parallax, proper motion, and quadratic time effect.* Since the recognition and evaluation of the instrumental equation in right-ascension over the interval 1942–1948, three solutions were published (van de Kamp, 1975, 1977a, 1979), all based on measurements made on the Grant machine (Chapter 2).

We now present the most recent 1980–analysis, based on all Sproul material, covering the 1916–1919; 1938–1979 interval, and using the three-star background. Beginning 1938 the star has been observed on an average of nearly thirty nights each observing season, which runs from the middle of March till the middle of September. The 1916–1919 parallax series, though limited in number of plates, was included because of its temporal value, in determining the quadratic time effect. A general least squares solution was made for proper motion, quadratic term and parallax using the formulae

$$\begin{aligned} X &= c_X + \mu_X t + q_X t^2 + \pi P_\alpha, \\ Y &= c_Y + \mu_Y t + q_Y t^2 + \pi P_\delta, \end{aligned} \tag{5.1}$$

where t is counted from 1950.0.

The value of P_α range from + 1.00 to −1.00, those of P_δ from about −0.05 mid March, up to +0.47 in June and down to +0.10 mid September.

The X-values over the interval 1942–1948 were corrected for instrumental equation by adding $+2.8\mu$ (Chapter 3). The following results based on plates taken on 1165 nights are obtained from a combined solution of the two coordinates (reduced to seconds of arc; scale value 1 mm = 18″87):

$$\begin{aligned} \mu_X &= -0''8037 \pm 0''0001, & \pi &= +0''5436 \pm 0''0009, \\ \mu_Y &= +10''2756 \pm 0''0001, & \pi_x - \pi_y &= +0''0243 \pm 0''0055, \\ q_X &= +0''000054 \pm 0''000004, & pe1x &= \pm 1.25\mu = \pm 0''0236, \\ q_Y &= +0''000301 \pm 0''000004, & pe1y &= \pm 1.31\mu = \pm 0''0247. \end{aligned}$$

The difference between the parallax values obtained from the two coordinates separately is explained by the above mentioned time-of-night effect in declination (Chapter 3).

Because of the limited range in P_δ, the declination data yield a parallax value whose weight is only a minute fraction of that obtained from the right ascension data. Hence parallax determinations of Barnard's star from a combined solution and from right ascension only are virtually the same.

Using + 0″.0034 as the average value of the parallaxes of the reference stars for the reduction to absolute (van de Kamp, 1963a), a value of 0″.548 is obtained for the absolute parallax. Other recent parallax determinations include the Allegheny value 0″.5429 ± 0″.0035 (m.e.) and the Van Vleck value 0″.5461 ± 0″.0035 (m.e.) which reduced to absolute yield 0″.546 and 0″.549 respectively (Gatewood and Eichhorn, 1973). Gliese (1969) gives 0″.552 ± 0″.001 (p.e.). The very small errors should be regarded with reservation, because of possible systematic errors (Lippincott, 1971). A final absolute value of 0″.547 ± 0″.003 is adopted.

The quadratic time effect, i.e. half the observed acceleration has been discussed in Chapter 8.

(e) *Normal points and weights.* The remainders from the solution (Section d) were combined into yearly normal points, which were next reduced from heliocentric to barycentric coordinates (Chapter 9). Over the interval covered by the observations the total range of this reduction is 0″.0085 or 0.45μ. The corrections applied range from -0.17μ to $+0.24\mu$ in x (RA), and -0.11μ to $+0.11\mu$ in y (Decl.).

It is a well-known fact that the results from a least squares solution based on extended homogeneous material depend little on the assigned weights. Generally, if a change in weights alters the results appreciably, this may be considered a sign of inadequate data. The total night weights for the yearly normal points range from 15 to 164; these are 'old' weights (van de Kamp, 1945). The least squares solution based on individual nights yields a $pe1$ of 1.28μ. Any solution based on yearly normal points, involving orbital motion or not, yields an appreciably higher error indicated by $pe1_*$ (Lippincott, 1957, 1971; van de Kamp, 1975). We attribute this to an *annual* error, varying from year to year, with a probable value pe_a. For a yearly normal point of total night weight W we thus have in either coordinate

$$(pe1_*)^2 = (pe1)^2 + W(pe_a)^2. \tag{16.1}$$

We adopt a value of $pe_a = 0.15\mu$ (van de Kamp, 1975). This implies a reduction from total annual night weight W to effective annual weight W_*

according to the relation

$$\frac{W_*}{W} = \frac{(pe1)^2}{(pe1)^2 + W(pe_a)^2} \tag{16.2}$$

or in the present case

$$W_* = \frac{W}{1 + 0.016W} \,. \tag{16.3}$$

The range in yearly total night weights, 15 to 164, is now reduced to a range from 10 to 45, with an average total reduced night weight of 30.

This concept of reduced *annual weight*, another illustration of the law of diminishing returns, throws light on the optimum total night weight worth striving for in any one year. For $pe1 = 1.2\mu$ and $pe_a = 0.15\mu$ we have the following relation

W	W_*
10	9
20	15
50	28
100	38

In other words by increasing the total night weight W five fold, from 10 to 50, we attain a three fold increase in effective annual weight W_*. By increasing the total annual night weight from 50 to 100, the additional increase in effective annual weight is only one third. Different values of the annual error will lead to different conclusions; there also remains the difference in accuracy between the two coordinates; an analogous situation was noted long ago for transit observations (Kapteyn, 1922). No harm ever can be done by adding more material, which in any case is always desirable for studying errors depending on the time of night (or season). In practise this may be done only for a limited number of stars. For the moment a conservative tentative rule might be for Barnard's star to aim at a minimum total annual night weight of 50, obtained on some 30 nights.

(f) *Orbital solutions.* Whatever the merits and limitations of the present analysis are, we wish to point out the striking, sustained continuity in the observed positions after correcting the RA data over the interval 1942–48, as represented by the annual normal points of the remainders (Section (d), Figure 16.6), particularly in right ascension beginning in 1949. Whether due

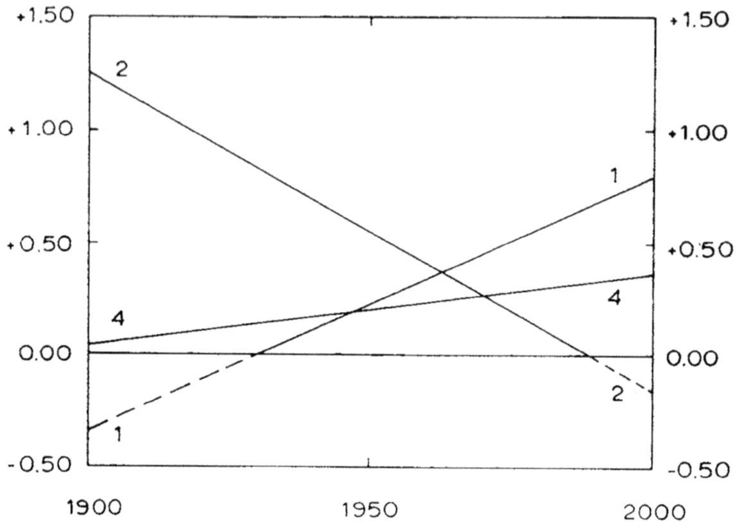

Fig. 16.5. Barnard's star. Dependence 'paths' of reference stars in the three-star configuration 124.

to cosmic or instrumental behavior, such continuity indicates the following: (1) consistent behavior of the telescope and measuring machine; (2) the value of strong material resulting from a large number of exposures obtained on many plates over as many nights as possible; (3) the significance of continuous temporal extent of the observations, in the present case every year for over four decades.

The pattern of the remainders for the yearly normal points over the interval 1938–79 can be represented by a combined 'short' and 'long' periodicity, each with a semi-amplitude close to half a micron or 0".01. Simultaneous solutions were made with the short period P_1 ranging from 11 to 14 yr, and the long period P_2 from 17 to 24 yr. The orbits were assumed to be circular.

The annual normal points were represented by the following formulae:

$$\begin{aligned} &\text{in } x \quad \Delta c_X + \Delta\mu_X + (B_1)x_1 + (G_1)y_1 + (B_2)x_2 + (G_2)y_2, \\ &\text{in } y \quad \Delta c_Y + \Delta\mu_Y + (A_1)x_1 + (F_1)y_1 + (A_2)x_2 + (F_2)y_2, \end{aligned} \qquad (16.4)$$

where (B), (G), (A), and (F) are the geometric elements (modified Thiele-Innes constants) of the perturbation (Chapter 13). The circular rectangular coordinates in unit orbit, $x = \cos M$, $y = \sin M$, and the mean anomaly M are counted from the arbitrary zero epoch 1950.00. After careful consideration the following preferred periods were adopted for further analysis

$$P_1 = 13.5 \text{ yr},$$

$$P_2 = 19.0 \text{ yr}.$$

The least squares solution from the 42 annual normal points over the interval 1938–79, with the preferred periods, yields the following values for the geometric elements and for the pel_*

	'short' period			'long' period	
P	13.5 yr			19.0 yr	
(B)	$+0.33$	$\pm 0.06\mu$		-0.19	$\pm 0.06\mu$
(G)	$+\ 9$	6		$+\ 4$	6
(A)	$+\ 16$	6		$-\ 34$	6
(F)	$+\ 19$	6		$-\ 24$	6
	$pel_{*x} = \pm 1.38\mu$		pel_* (average of x and y) $=$	$\pm 1.46\mu$	
	$pel_{*y} = \pm 1.55\mu$				

For the average reduced annual weight 30 of the yearly normal points the corresponding probable error is 0.27μ or $0''.005$.

The geometric elements yield the orbital elements α, i, and Ω and, combined

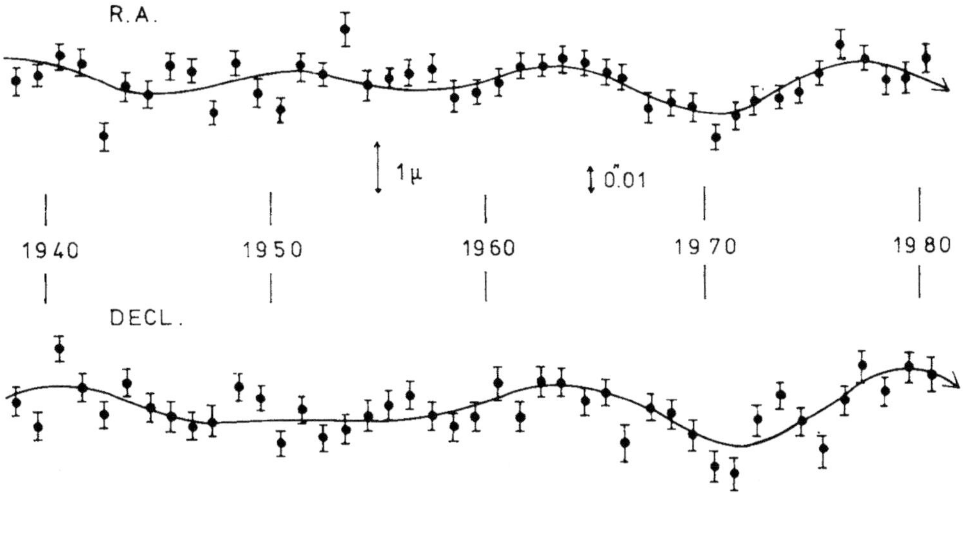

Fig. 16.6. Barnard's star. Yearly normal points and calculated orbital displacement curves in RA and Decl over the interval 1938–1980 from Sproul photographs obtained on 1165 nights. The recently completed normal point based on 32 nights in 1980, is close to the ephemeris (residual $+\ 0.4\mu$ in RA, 0.0μ in Decl.).

with the period P, the orbital constant $\alpha P^{-2/3}$:

	'short' period	'long' period
α	$0.41 \pm 0.06\mu = 0.0140$ AU	$0.44 \pm 0.06\mu = 0.0152$ AU
i	$107° \pm 5°$	$108° \pm 5°$
☊	$56° \pm 5°$	$20° \pm 5°$
$\alpha P^{-2/3}$	0.0025	0.0021
	$1\mu = 0''.019 = 0.0345$ AU	

The relative inclination I of the two orbits is given by

$$\cos I = \cos i_1 \cos i_2 + \sin i_1 \sin i_2 \cos (☊_1 - ☊_2). \tag{16.5}$$

While i_1 and i_2 are uniquely defined, the values of $☊_1$, or $☊_2$, may be changed by 180°. Hence the second term on the right hand side of the equation has a

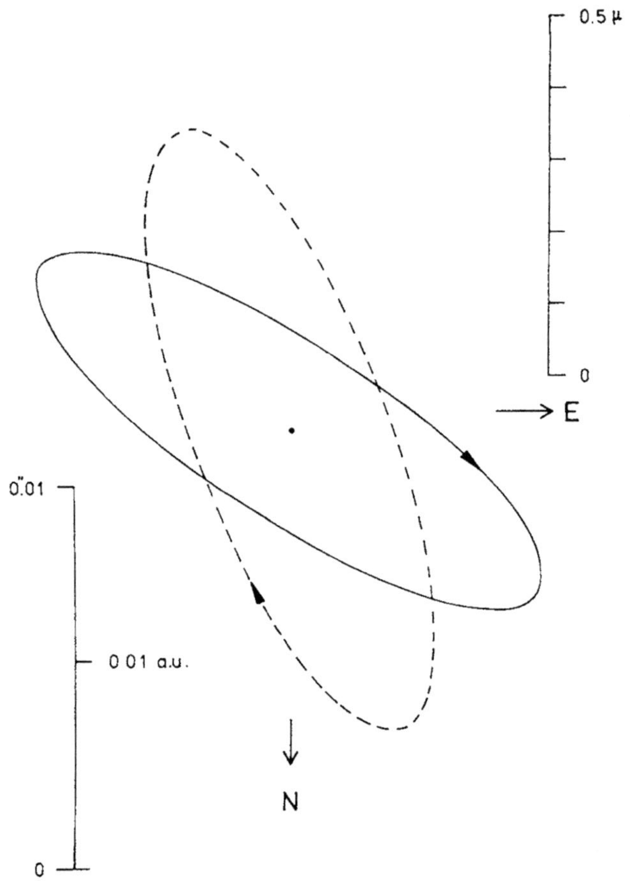

Fig. 16.7. Barnard's star. Calculated apparent perturbation orbits. *Solid curve:* period 13.5 yr, linear radius 0.0140 AU; *broken curve*: period 19.0 yr, linear radius 0.0152 AU.

double sign and two different values of I are found, in the present case 34° and 130°. The angle between the two orbital planes could be appreciably altered by permissible adjustments to the geometric orbital elements. We have preferred to 'let the observations speak for themselves', and have not tried to accommodate *a priori* concepts such as near co-planarity or co-revolution.

The position angle Θ of the line of intersection of the two orbits is given by

$$\tan \Theta = \frac{\cot i_1 \sin \Omega_1 - \cot i_2 \sin \Omega_2}{\cot i_1 \cos \Omega_1 - \cot i_2 \cos \Omega_2} \tag{16.6}$$

which in the present case yields $\Theta = 133°$.

(g) *Dynamical interpretation.* The very small orbital constants $\alpha P^{-2/3}$ indicate small values for the masses of the unseen companions attributed to the two perturbation orbits and we thus may use the simple formula

$$\mathcal{M}_B = \alpha P^{-2/3} \mathcal{M}_A^{2/3}. \tag{13.19}$$

An adopted mass of Barnard's star is based on the near equality in absolute magnitude ($M_v = 13.2$) and infra-red color index with those of Ross 614 A ($M_v = 13.3$) and of Krüger 60 B ($M_v = 13.2$) (Lippincott and Hershey, 1972; Lippincott, 1953; Wanner, 1967), for which the mass values are 0.11 $\mathcal{M}_\odot$ and 0.16 $\mathcal{M}_\odot$, respectively. For an adopted mass of 0.14 $\mathcal{M}_\odot$ for Barnard's star, the 'short' and 'long' orbits of the two companions around Barnard's star are found to have radii of 2.95 AU and 3.70 AU (13.20) and greatest elongations of 1.″6 and 2.″0, respectively. No trace of companions has been detected on the well over ten thousand exposures, and we assume that they contribute no light to the measured images. The results are summarized:

	'short' orbit	'long' orbit
Radius of relative orbit	2.95 AU = 1.″6	3.70 AU = 2.″0
Radius of perturbation orbit	0.0140 AU = 0.″0077	0.0152 AU = 0.″0083
Mass ratio companion to primary	0.0047	0.0041
Mass of companion	0.00066 $\mathcal{M}_\odot$ = 0.7 $\mathcal{M}_♃$	0.00057 $\mathcal{M}_\odot$ = 0.6 $\mathcal{M}_♃$

For each companion there is a phase difference of 180° between the position in the perturbation orbit and in the relative orbit of the companion around the primary (Figures 16.7 and 16.8).

The current results differ slightly from earlier ones (van de Kamp, 1975, 1977a, 1979, 1980). The two latter results covered the interval 1950–1978 only, permitting the inclusion of reference star 5, which, because of the large proper motion of Barnard's star, does not appear on earlier plates. Although in the forseeable future the four star combination appears to be ideal, at present the inclusion of star 5 is of minimal significance, its dependence ranging from–0.02 in 1950 to +0.12 in 1980 and the present (1980) geometric error (Chapter 4, Section (d)) of a reduced position averaging only 6% more for the three star background. The instrumental equation in RA over the interval 1942–48 appears now well established; using the three star background permits a continuous plate series, covering 42 consecutive years permitting for the first time a fair separation of the short period, covered three times, and the long period, covered twice.

A tabulation of the results of the three earlier Sproul analyses, in which the 1942–1948 material was omitted, with the present results based on all material is given below:

Interval	Period		Companion mass relative to Jupiter		Reference van de Kamp
	'short'	'long'			
1950–1974	11.5 yr	22.0 yr	$1.0\mathcal{M}_{\jupiter}$	$0.4\mathcal{M}_{\jupiter}$	(1975)
1938–41; 1949–75	11.7	18.5	0.9	0.4	(1977a)
1950–1978	11.7	20.0	0.8	0.4	(1979, 1980)
1916–19; 1938–79	13.5	19.0	0.7	0.6	(1980, present)

The principal difference between the current (3-star background, interval 1916–19; 1938–79) and last years' interpretation, (4-star background, interval 1950–1978) is a greater amplitude of the 'long' orbit, resulting in an increase in the mass of the companion from $0.4\,\mathcal{M}_{\jupiter}$ to $0.6\,\mathcal{M}_{\jupiter}$; the mass of the companion in the short orbit is reduced from $0.8\,\mathcal{M}_{\jupiter}$ to $0.7\,\mathcal{M}_{\jupiter}$. Both masses have probable errors of $\pm 0.1\mathcal{M}_{\jupiter}$. Considering the various uncertainties, the essential interpretation remains the same: two planetary companions with masses somewhat below that of Jupiter. They would be too faint to be detected by direct observation with existing instruments, the expected apparent magnitudes being about 30.

The next few years should witness marked improvement for the amplitudes and periods of the two component perturbations and hence for the masses of the two companions.

(h) *Possible influence of reference stars.* To test the possibility that the observed perturbation might be a reflex of orbital behavior of a reference star, solutions were made for parallax and proper motion of the reference stars 1, 2, 4, and 5, with respect to their mean. The relative parallaxes were found to be below 0″001 with probable errors less than 0.1μ or 0″002. This by itself would appear to preclude a measurable perturbation for any of the reference stars. We checked for any non-linear proper motion within the interval 1970–1978, observationally well covered, in which the orbital effect for Barnard's star increases $+0.9\mu$ in RA and $+1.2\mu$ in Decl. If this were to be attributed to a reference star, the latter should show an orbital effect,

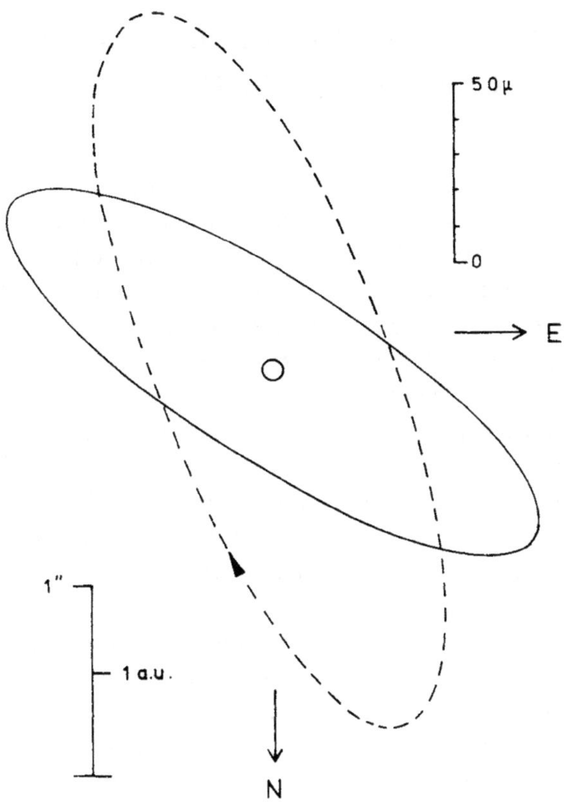

Fig. 16.8. Barnard's star. Calculated apparent orbits of invisible companions. *Solid curve:* period 13.5 yr, linear radius 2.95 AU; *broken curve:* period 19.0 yr, linear radius 3.70 AU.

decreasing by the same amount, divided by the dependence of the reference star. The expected decrease over the interval 1950–1978 would be about 2μ for star 1 and double that amount for star 2 or star 4. There is no indication whatsoever of such large effects for any of the reference stars; on the contrary their yearly normal points indicate uniform proper motions within 0.2μ.

CHAPTER 17

LONG-PERIOD ECLIPSING BINARIES: VV CEPHEI AND EPSILON AURIGAE

(a) *Apparent orbit vs annual parallax.* A determination of the astrometric orbit of a spectroscopic binary is of interest for its own sake but also because it may permit the determination of a precise (absolute) parallax if the linear dimension of the orbit is sufficiently large. This situation turns out to be of particular interest for the two eclipsing spectroscopic binaries of known longest periods, 20.4 and 27.08 yr, for the naked-eye stars VV Cephei and Epsilon Aurigae respectively. In each of these systems the brighter component is an extreme supergiant. The primary of VV Cephei may well be the largest known red supergiant star, so extended that with the Sun at its center the orbit of Mars would easily be contained inside it. The primary of Epsilon Aurigae has a diameter exceeding that of the Earth's orbit, and may well be the F star of highest known luminosity; its massive invisible companion remains a mysterious object.

Both these binary systems have been extensively studied, spectroscopically and photometrically. Their spectra have led to estimates of their luminosities, but in principle the luminosities should be evaluated directly from the apparent magnitudes and precise values of the distance from us. How can the latter be obtained with adequate accuracy? A conventional annual parallax determination may at best yield an accuracy of $\pm 0\overset{''}{.}002$, or even $\pm 0\overset{''}{.}001$ if a supreme effort is made. Small as such an error may seem it is of the same order of size as the parallax itself for distant supergiants. But another, geometric approach exists.

An annual parallax determination consists of measuring the unforshortened angular value of one astronomical unit as it would be seen from the star, or which amounts to the same, as it would appear to us at the location of the star (Chapter 5). However, the orbit of a long period eclipsing binary is much larger than the earth's orbit, say 10 times as large, and its linear dimensions may be known from spectroscopic and photometric observations (the latter also providing the orbital inclination). If the orbit has been sufficiently covered by long focus photography, its angular dimensions may be obtained. Dividing the angular (seconds of arc) by the linear size (astronomical units) then gives us a much more precise value of the parallax than is possible with the conventional annual parallax method. Greater parallactic leverage is reached, as if observations were made from the orbit of Saturn or Uranus as a baseline. The apparent orbit may be that of the primary component only, when the secondary is without any appreciable luminosity (as in the case of Epsilon Aurigae), or we may measure the photocentric orbit revealed by the blended light of the two components (as in the case for VV Cephei). In both cases these orbits may be analyzed as a 'perturbation' caused by the secondary.

(b) *Concept of orbital parallax.* The analysis of the stellar path is obtained by representing the observations by the following formulae:

$$X = c_X + \mu_X t + \pi P_\alpha + (B)x + (G)y,$$
$$Y = c_Y + \mu_Y t + \pi P_\delta + (A)x + (F)y. \tag{17.1}$$

The right-hand side of these equations represents the proper motion of the barycenter, the parallax and the star's own apparent orbit. Quadratic time effects (Chapters 5, 8) are entirely negligible for these distant objects of very small proper motion. The elliptical rectangular coordinates x and y may be calculated from the dynamical elements P, e, T determined photometrically and/or spectroscopically. The values of the geometric elements (B), (G), (A), (F) determined from sufficient astrometric material yield the scale α and the orientation elements i, ω, and Ω (Chapter 13). The semi-axis major α expressed in angular measure refers to the photocenter or to the primary if the secondary is truly dark. The values of i and (to a lesser extent ω) if in good agreement with the photometrically determined values may be considered an argument for the reality, and precision, of the astrometric orbit; a value for Ω can be obtained only from an astrometric study. While the value of π must be reduced to absolute to obtain the absolute value p of the parallax, the latter named *orbital parallax* p_0 may now be obtained directly from the relation

$$p_0'' = \frac{\alpha''}{\alpha}, \tag{17.2}$$

where α'' is the observed semi-axis major in angular measure and α the linear value in astronomical units obtained from the spectroscopically determined value of $\alpha \sin i$, using the value of i determined photometrically or astrometrically. The values of α and π in seconds of arc are likely to have the same accuracy. Therefore if α is of the order of say 10 astronomical units, the error of the value of the absolute orbital parallax obtained by dividing α'' by α is obviously much smaller than the error of the parallax, obtained from the direct determination of π, which moreover has to be reduced to absolute.

In 1938 both VV Cephei and Epsilon Aurigae were placed on the astrometric program of the Sproul Observatory refractor. Our hope was that the apparent orbits of both systems would be large enough to be revealed, and thereby yield parallaxes by the method just described. After four decades of intensive observations, this expectation has been fulfilled for both binaries; continued observations of each system remain of course desirable.

(c) *VV Cephei* (van de Kamp, 1977b, 1978b). Sergei Gaposchkin (1938) discovered that VV Cephei is an eclipsing binary, although it was already known to be a variable star and a spectroscopic binary. The system consists of a slightly variable M supergiant of average apparent visual magnitude 5.25, and a B9 companion of visual magnitude 6.97 (Fredrick, 1960). From the available spectroscopic and photometric data (Wright, 1970), we adopt 25 AU for the radius of the relative orbit (assumed circular) of the two components. The radii of the orbits of the M and B components around their barycenter are 13 and 12 AU, respectively, and their masses are 18.3 and 19.7 $\mathcal{M}_{\odot}$. Because of the huge size of the M star, it hides the B star for about 600 days during each primary eclipse. As observed in blue light, primary eclipse is about 0.6 mag. deep, but in visual light it is too shallow to be detected with certainty except with a photoelectric photometer. Secondary eclipse, which occurs when the small hot B star passes in front of the large cool M star, is too shallow to observe. Photographs taken with the 61-cm Sproul refractor do not resolve the binary, but yield a blended image which describes an orbit whose size is 0.35 times that of the relative orbit, this ratio being known from the relative brightness ($\Delta m = 1.72$, $\beta = 0.17$) of the components. The radius of this photocentric orbit, $0.35 \times 25 = 8.75$ AU, is therefore much larger than the annual parallax, which corresponds to 1 AU.

Measurement of 2077 Sproul plates taken on 579 nights over the interval 1938 to 1976 confirm the instrumental equation in RA over the interval 1942–1948 (none in Decl.). A correction of $+2.4\mu$ is required to provide continuity with earlier, but primarily with later material.

Both parallax and apparent orbit are very small, the period of the apparent orbit is much larger than the one year period of the parallactic orbit. We are therefore justified to make separate solutions for parallax and for apparent orbit. A solution for parallax only, yields

$$\pi = 0''.0022 \pm 0''.0011$$

or $+0''.0034$ reduced to absolute (Fredrick, 1960). The yearly mean remainders clearly reveal an orbital pattern with a total amplitude of over one micron and a period of about two decades, consistent with the spectroscopic and photometric period. We have chosen to make an orbital solution limiting ourselves to the material beginning with 1950 (1479 plates on 428 nights), because of the superior quality of the observational material since 1950.

Adopting $P = 20.4$ yr (Hutchings and Wright, 1971) but a circular orbit ($e = 0$) which gives a better representation of the epoch of mid eclipse, we obtain

$$\alpha = 0''.0120 \pm 0''.0013,$$
$$\Omega = 147^\circ \pm 3^\circ,$$
$$i = 74^\circ \pm 3^\circ.$$

Equating this value of α with the linear value 8.75 AU (17.2) we obtain $0''.0014 \pm 0''.0001^5$ for the 'orbital' parallax, corresponding to a distance of 700 parsec or 2300 light years. The computed error is a lower limit, primarily due to unrecognized errors and in the blending factor β. Nevertheless, note the extreme precision of the orbital parallax value, compared with the error $\pm 0''.0011$ of the direct determination of the relative annual parallax.

With this new orbital parallax value, the absolute visual magnitudes of the

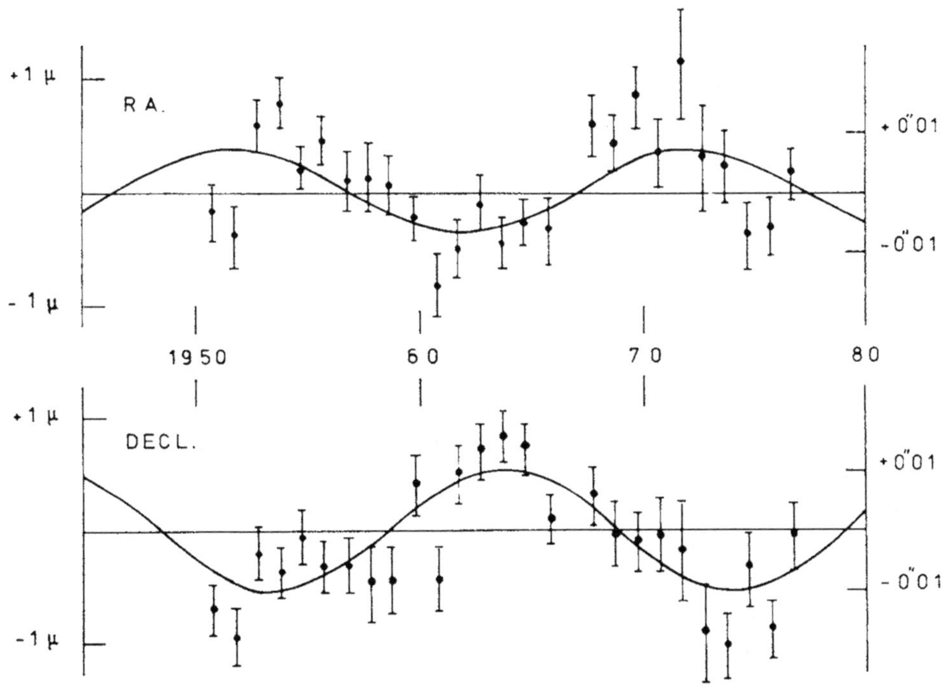

Fig. 17.1. VV Cephei. Yearly normal points and calculated orbital displacement curves in RA and Decl.

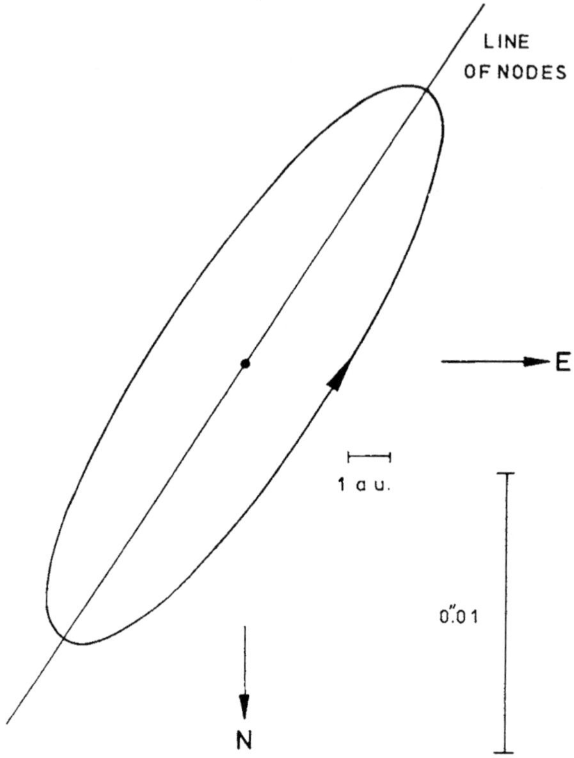

Fig. 17.2. VV Cephei. Calculated apparent orbit.

M and B components are −4.0 and −2.3, respectively with probable errors of ±0.15 mag. If we assume an interstellar dimming of 0.3 magnitudes, the absolute magnitudes are −4.3 and −2.6, corresponding to visual luminosities for the M and B components of 5800 and 1200 suns, respectively. The maximum angular separation of the two components will be about 0.″06 in the early 1980's. This is too close for visual resolution in any existing telescope, but the new technique of speckle interferometry may permit resolution.

If VV Cephei were as close to us as Alpha Centrauri, it would be easily the most remarkable double star in the heavens. It would be observable even in day time, as a red disk a second of arc in diameter, of magnitude −8.5 (nearly as bright as the quarter moon), with a blue star of −6.5 (considerably outshining Venus) less than 30″ away.

(d) *Epsilon Aurigae* (van de Kamp, 1978a, b). The variability of Epsilon Aurigae has been known since the middle of the 19th century, for this is a relatively bright star whose eclipses, 0.7 mag. deep and lasting about 700 days, are easily observed with the unaided eye. (The next eclipse begins in July 1982.) This binary with a period of 27.08 yr consists of a supergiant F star of visual magnitude 2.96 and an invisible companion, whose nature has been the subject of many studies and speculations. It has been interpreted as a virtually transparent disk of dust and gas, strongly concentrated around a central star that may be a hot B type one. In another model, the unseen secondary is a cool disk of solid particles surrounding a collapsed star. The two components have masses of 15.5 and 13.7 $\mathcal{M}_\odot$. From the spectroscopic orbit determination by Wright (1970), we can adopt 13.2 AU as the radius of the orbit (assumed circular) of the F star around the barycenter.

Measurement of 1090 plates taken on 301 night over the interval 1939–1977 indicate an instrumental equation in RA for this blue star over the interval 1942–1948. A correction of -1.0μ was adopted to provide continuity with earlier and later material.

As in the case of VV Cephei a circular oribt yields a better representation of the epoch of mideclipse. Adopting $P = 27.08$, a combined analysis for

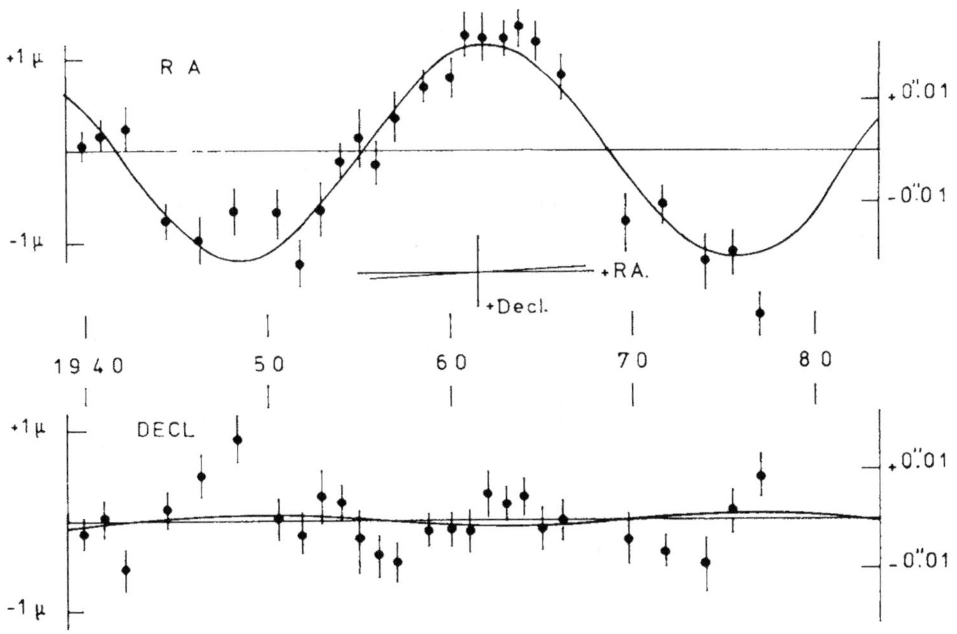

Fig. 17.3. Epsilon Aurigae. Yearly normal points and calculated orbital displacement curves in RA and Decl. Insert: apparent 'linear' orbit.

parallax and orbit yields

$$\begin{aligned} \pi &= -0\overset{\prime\prime}{.}0019 \pm 0\overset{\prime\prime}{.}0010, \\ \alpha &= +0\overset{\prime\prime}{.}0227 \pm 0\overset{\prime\prime}{.}0010, \\ \Omega &= 92^\circ \pm 3^\circ \\ i &= 89^\circ \pm 3^\circ. \end{aligned} \tag{17.3}$$

Equating the angular value of α with the linear value 13.2 AU, an orbital parallax $+0\overset{\prime\prime}{.}00172 \pm 0\overset{\prime\prime}{.}00008$ is found, corresponding to a distance of 580 parsec or 1900 light years. Again note the extreme precision of the orbital parallax as compared with the value $\pm 0\overset{\prime\prime}{.}0010$ for the error of the relative annual parallax. The precision is even better, proportionally, than for VV Cephei because of the larger orbit observed for Epsilon Aurigae.

The orbital parallax yield an absolute visual magnitude of -5.9 ± 0.1 for the visible F component. If we allow a total visual absorption of 0.84 (Morris, 1963) the absolute visual magnitude is found to be -6.7, or a visual luminosity of about 40000 suns. There is little prospect of resolving this binary by any existing technique since the maximum separation, predicted for 1989, is only $0\overset{\prime\prime}{.}02$, and the companion must be very faint.

(e) *Summary*. For both VV Cephei and Epsilon Aurigae, long period eclipsing spectroscopic binaries, the apparent astrometric orbits have now been measured and yield accurate values for the orbital parallax of these systems, far more accurate than the conventional annual parallax determinations. Confidence in the reality of the astrometric results for both stars is strengthened by the high orbital inclinations (74° and 89°, respectively) from the photographic measurements. Also, the times of mideclipse as found from the astrometric and photometric analyses are within one year in each case.

As mentioned before, neither spectroscopic nor photometric observations of an eclipsing binary can determine the direction of the line of nodes, that is, the position angle of the line of intersection that the orbit plane makes with the 'plane of the sky'. But the astrometric data can, and give the position angle of the line of nodes as 147° for VV Cephei and 92° for Epsilon Aurigae.

I am indebted to the late Joseph Ashbrook for numerous editorial improvements of this presentation.

CHAPTER 18

EPILOGUE ATTAINABLE ACCURACY. SUBSTELLAR AND PLANETARY DETECTABILITY

(a) *Review*. Studies of stellar paths by the technique and methods of long-focus photographic astrometry, begun shortly after the turn of the century, are being pursued with enthusiasm and intense efforts. Increased precision has been due to the dramatic increase in the speed of photographic emulsions, improved operation of older telescopes, the construction of new specially designed astrometric telescopes, and, last but not least, the introduction of high speed precision measuring machines. The results have contributed to our knowledge of motions, distances, orbital motions, masses and luminosities of stars. The past several decades moreover have witnessed a marked development in the field of perturbations in stellar paths, leading to the discovery and subsequent study of unseen companions. These have proven to be mostly bona-fide stars, but also include substellar and even planet-like objects.

(b) *Separating small perturbations from random errors*. The subject of attainable accuracy and sustained astrometric precision remains important, particularly for the study of small perturbations. While more than a score well established unresolved astrometric orbits and several marginal results have been found, numerous stars have yielded no measurable deviations, even from intensive observations extending over several decades. In no case of course, may such evidence be taken to indicate the absence of any perturbation. It may simply be a case of the perturbation being so small that it is well below the threshold value of observational accuracy, possibly because of short period, or the period may be so long that no measurable 'curvature' has yet occurred. In either case, final judgment has to be postponed.

We quote Bessel (1845), after his classical discovery of the perturbations in the proper motions of Sirius and Procyon.

"But light is no real property of mass. The existence of numberless visible stars can prove nothing against the existence of numberless invisible ones".

"For even, if a change of motion can up to the present time, be proved in only two cases, yet will all other cases be rendered thereby liable to suspicion, and it will be equally difficult by observations, to free other proper motions from the suspicion of change, and to get such knowledge of the changes as to admit of its amount being calculated".

Bessel clearly anticipated the present situation. The absence of a measurable perturbation is no argument against the existence of a perturbation. We need not be discouraged. For those cases where no measurable perturbation has (yet) been found, we can at least put an upper limit to the mass(es) of any unseen companion(s) for a range of periods. But also, the many cases of no perturbation, i.e. a constant proper motion, serve as a check, even support of the reality of what for other stars appear to be well established perturbations of larger amplitude. The critical boundary is something like a total spread in annual residual positions (after allowing for parallax, proper motion and acceleration) of one micron (about 0".02), depending on the number of periods covered.

The following consideration is illuminating. Intensive observations may yield at best a probable error for an annual normal point of 0.1μ or 0".002. Assume now that a star would have a perturbation, well covered by the observations, with a semi-axis major 0".01 (in each coordinate). The spread of the orbital displacement may be approximatively represented by an error distribution with a probable error of $0''.02/\pi$ or 0".0064. Combining this with a yearly observational error of 0".002 yields a probable error of 0".0067. A larger yearly observational error of 0".005 would result in an combined error of 0".008. This demonstrates the problem and difficulty of extracting perturbations of small amplitude from the observations. But while the observational errors are *random*, the orbital deviations follow a *systematic periodic* pattern.

An illustration. For Barnard's star (Chapter 16), the average residual for a normal point over the interval 1938–1979 is 0.29μ or 0".0055, after allowing for two perturbations with periods 13.5 and 19.0 yr, and semi-axes major of 0.41μ and 0.44μ, or close to 0".008. If the perturbations are not allowed for, the average residual for a normal point goes up to 0.37μ or 0".007. The difference between 0".007 and 0".0055 could have been ascribed to a random spread with an average value $\sqrt{(0''.007)^2 - (0''.0055)^2}$ or slightly over 0".004.

Rather than ignoring this small difference we have recognized the *systematic* behavior and have chosen to represent this spread by two perturbations.

Even if one would not recognize or accept the reality of the perturbations, the average residual of 0".007 would still make a valid argument for the essential stability, at the fractional micron level, of the Sproul telescope year in, year out, over the entire 42 yr interval.

(c) *Long-range telescope stability.* To illustrate this further we quote from a recent study by Hershey *et al.* (1980) of three forty-year intensive Sproul plate series, in which they discuss telescope stability and in particular planetary detection capability and probability. Hershey *et al.* studied plate series on three nearby red dwarf stars: all plates were taken with the Sproul refractor and measured on the two coordinate Grant machine at the Sproul Observatory. Instrumental equation in RA over the interval 1941–1949 was allowed for.

Starfield	Interval	Number of nights
Wolf 294	1938–1978	195
Groomb.1618	1939–1977	183
Ross 128	1943–1978	122

The residuals of the measured positions were tested for the presence of perturbations of short periods, 0.6 yr and longer. There is no conclusive indication for any measurable perturbation for any one of the three stars. The telescope evidently has shown sub-micron or milli-seconds stability across four decades for these three, and many other fields, apart from the short interval 1941–1949 in RA. The errors are slightly smaller in Decl. than in RA.

For the last 20 years, where the yearly weight generally has been greater, the yearly normal points in Decl. (average of 3 yr) have a probable error of below 0.1μ or two milliseconds of arc. Similar accuracy has been found in studies of average Sproul residuals by Lippincott (1971). The stability of the telescope is remarkable, especially so considering that it was designed at the beginning of the century as a visual instrument without the stringent requirements of present day photographic astrometry in mind.

The current long range accuracy of the Sproul telescope and Grant measuring machine are especially evident on plate series from the late 1960's

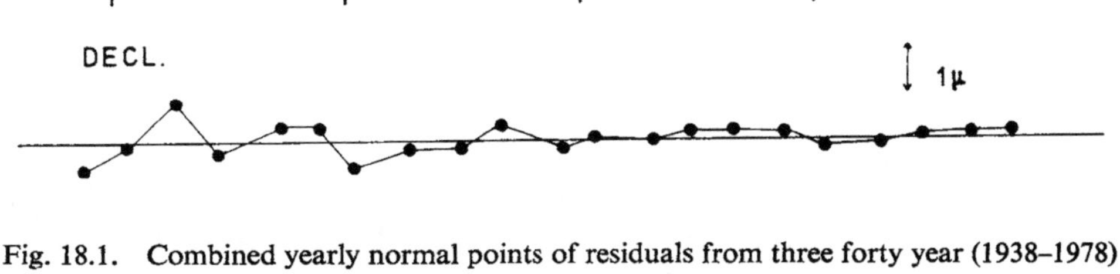

Fig. 18.1. Combined yearly normal points of residuals from three forty year (1938–1978) Sproul solutions for parallax and proper motion. The short interval 1941.8–1949.2 was adjusted for instrumental equation in RA.

and throughout the decade of the 1970's. The mechanical and electronic renovation of the telescope in 1967 (Wanner, 1968), the increased quality of photographic emulsions, and the accuracy of the measuring machine have combined to yield a growing number of recent (parallax) plate series of approximately 10 years' length which yield probable errors of unit weight (one plate with two exposures) under 1μ and as low as 0.6μ. The yearly mean residuals based on about a dozen plates often have probable errors less than 0.2 or 0.3μ (2000–3000 Å). A series of moderate size (70 nights) can produce a formal (internal) probable error for the parallax of 0."002 (1000 Å) or less (Chapter 5).

In general the Sproul telescope has provided surveillance over four decades at a level of 0."010 for star fields with intensive plate coverage. This level of stability implies powerful discovery capability for perturbations with periods from one to four decades.

(d) *Substellar and planetary detection capability and probability.* The telescope reliability for stars without perturbation serves as a control and supports the reality of small amplitudes indicating the presence of substellar companions. This stability for the Sproul telescope is also substantiated by statistical considerations of a total of 23 Sproul astrometric orbits compiled by Lippincott (1978) plus two additional orbits (Lippincott, 1979). If through unlikely instrumental behavior, the telescope were generating spurious orbits of small amplitude there would be an increasing abundance of perturbation orbits below 0."02 amplitude; this proves not to be the case. If the telescope had a long term periodic change or a series of discrete changes of a quasi-periodic nature, there would be an excess of particular periods; again this is not the case. The statistics of the Sproul astrometric orbits are very healthy and lend additional support for the long-range stability of the Sproul 61 cm refractor (Hershey *et al.*, 1980).

Upper limits can be set on the mass of unseen companions by fitting the residuals from a proper motion and parallax solution, with orbits over a wide range of periods. The maximum mass $\mathcal{M}_B$ of a single unseen companion with a mass much below the mass $\mathcal{M}_A$ of the primary is very approximately

$$\mathcal{M}_B = \alpha P^{-2/3} \mathcal{M}_A^{2/3}, \qquad (13.19)$$

where α is the semi-axis major of the perturbation expressed in AU. Assuming various values for the period P, least squares solutions for the residuals yield various value of α. And for a range of values of $\mathcal{M}_A$, maximum values of the mass of the unseen companion can be computed.

For the three stars studied by Hershey *et al.* the detection capability reaches toward one to several Jupiter masses at periods from four down to one or two decades.

The question of long-range telescope stability and the detectability of unseen companions of low mass are closely related. From a study of visual and spectroscopic binaries, Abt and Levy (1976) and Abt (1977, 1979) find that secondary companions of F and early G type stars decrease in number as the cube root of their mass for periods under 100 yr. Abt finds it reasonable to suppose that this law also holds for companions of lower main sequence stars. Detectable companion masses of substellar nature may be taken to range from 0.001 $\mathcal{M}_\odot$ as a lower detectable limit to 0.06 $\mathcal{M}_\odot$ as a lower limit for a *bona fide* star. Assuming primary masses of 0.3 $\mathcal{M}_\odot$ as a representative mass for the great majority of lower main sequence stars in our neighborhood (say within 5 parsec) the number of unseen companions between 0.001 $\mathcal{M}_\odot$ and 0.06 $\mathcal{M}_\odot$ would be somewhat under one half of the total number of companions.

Considerations of the expected distribution of periods show that roughly one – tenth of the nearby stars, within 5 parsec, might be expected to have astrometrically detectable sub-stellar companions. This would amount to a few objects only, for the nearby stars which have had long term observations. And some may actually have been found (Barnard's star, CC1228, BD + 68° 946, BD + 43° 4305, Stein 2051 A)!

That several stars have not yielded any measurable perturbations, does not preclude the existence of a large number of planets with Jupiter-like or smaller masses, with short periods and therefore amplitudes too small for astrometric detection. Longer intervals of observation and/or new methods are needed. However new instrumentation cannot accelerate the passage of time. There always will be problematic cases of amplitudes near the limits of instrumental accuracy, regardless of how high the accuracy may be. Long-term stability at limiting levels of long-focus photographic accuracy cannot be established until long intervals of time have elapsed. Statistically, lower mass objects require longer time intervals of observation. A doubling of long-term accuracy still would require 15 yr to equal the same level of mass detection of a 40-year series. A Jupiter mass with a period of a century would be clearly detectable for nearby stars with existing telescopes and techniques if surveillance is continued for several decades.

REFERENCES

Abt, H. A.: 1977, *Scientific American* **236**, 96.

Abt, H. A.: 1979, *Astron. J.* **84**, 1591.

Abt, H. A. and Levy, S. G.: 1976, *Astrophys. J. Supplement* **30**, 273.

Adams, W. S. and Joy, A. H.: 1926, *Publ. Astron. Soc. Pacific* **38**, 122.

Alden, H. L.: 1924, *Astron. J.* **35**, 133.

Alden, H. L.: 1947, *Astron. J.* **52**, 138.

Åstrand, J. J.: 1890, *Hilfstafeln*, Wilhelm Engelmann, Leipzig.

Barnard, E. E.: 1916a, *Astron. J.* **29**, 181.

Barnard, E. E.: 1916b, *Pop. Astron.* **24**, 504.

Barnard, E. E.: 1916c, *Pop. Astron.* **24**, 586; *Publ. Am. Astron. Soc.* **3**, 191.

Bessel, F. W.: 1844, *Astron. Nachr.* **22**, 145.

Bessel, F. W.: 1845, *Monthly Notices Roy. Astron. Soc.* **6**, 136.

Bieger, G. S.: 1964, *Astron. J.* **69**, 812.

Bos, van den, W. H.: 1928, *Mémoires de l'Académie Royale des Sciences et des Lettres de Danemark*, Section d. Sc. 8me série. t. XII no 2.

Boss, B. *et al.*: 1937, *General Catalogue of 33342 Stars*, Vol. I, Appendix III.

Boss, L.: 1910, *Preliminary General Catalogue of 6188 Stars*, Appendix III.

Clemence, G. M.: 1953, *Astron. Papers of the American Ephemeris and Nautical Almanac* **13**, Part 4.

Crommelin, A. C. D.: 1916, *Monthly Notices Roy. Astron. Soc.* **77**, 4.

Deutsch, A. N.: 1978, *Pisma Astron. Zh. Akad. Nauk SSSR* m 4, No. 2.

Deutsch, A. N. and Orlova, O. N.: 1977, *Astron. Zh. Akad. Nauk SSSR* **54**, 327; also *Soviet Astron.* **21**, No. 2.

Feierman, B.: 1971, *Astron. J.* **76**, 73.

Franz, O. G.: 1958, *Astron. J.* **63**, 329.

Franz, O. G. and Mintz, B. F.: 1964, *Astron. Papers of the American Ephemeris and Nautical Almanac* **19**, Part 1.

Fredrick, L. W.: 1960, *Astron. J.* **65**, 628.

Gaposchkin, S.: 1938, *Harvard Circular*, No. 421.

Gasteyer, G.: 1954, *Astron. J.* **89**, 243.

Gatewood, G.: 1972, Ph.D. Dissertation.

Gatewood, G. and Eichhorn, H.: 1973, *Astron. J.* **78**, 769.

Gatewood, G. and Russell, J.: 1974, *Astron. J.* **79**, 815.

Gliese, W.: 1956, *Z. Astrophys.* **59**, 1.

Gliese, W.: 1969, *Veröff. Astron. Recheninstitut Heidelberg*, No. 22.

Gliese, W.: 1970, *Veröff. Astron. Recheninstitut Heidelberg*, No. 24.

Gliese, W.: 1978, in A. G. D. Philip and D. S. Hayes (eds.), 'The HR Diagram', *IAU Symp.* **80**, 79.

Greenstein, J.: 1972, *Astrophys. J.* **173**, 377.

Grossman, A. S.: 1970, *Astrophys. J.* **161**, 619.

Hall, Jr., R. G.: 1951, *Astron. J.* **55**, 215.

Harrington, R. S.: 1968, *Astron. J.* **73**, 508.
Harrington, R. S.: 1971, *Astron. J.* **76**, 931.
Harrington, R.S.: 1977, *Publ. Astron. Soc. Pacific* **89**, 244.
Harrington, R. S., Christy, J. W., and K. Aa. Strand: 1981, *Astron J.* **86**, 909.
Hershey, J. L.: 1973a, *Astron. J.* **78**, 421.
Hershey, J. L.: 1973b, *Astron. J.* **78**, 935.
Hershey, J. L.: 1975a, *Astron. J.* **80**, 662.
Hershey, J. L.: 1975b, *Image Processing Techniques in Astronomy*, D. Reidel Publ. Co., Dordrecht, Holland, p. 211.
Hershey, J. L.: 1977, *Astron. J.* **82**, 179.
Hershey, J. L.: 1978, *Astron. J.* **83**, 197.
Hershey, J. L.: Borgman, E. R., and Worth, M. D.: 1980, *Astrophys J.* **240**, 130.
Hertzsprung, E.: 1919, *Astron. Nachr.* **208**, 115.
Hertzsprung, E.: 1920, *Publ. Astrophys. Observatorium Potsdam* **24**, No. 75.
Hutchings, J. B. and Wright, K. O.: 1971, *Monthly Notices Roy. Astron. Soc.* **155**, 203.
Ianna, P. A.: 1965, *Vistas Astron.* **6**, 93.
Kamp, van de, P.: 1934, *Astron. J.* **44**, 74, 96.
Kamp, van de, P.: 1939, *Astron. J.* **48**, 21.
Kamp, van de, P.: 1943, *Pop. Astron.* **51**, 175.
Kamp. van de, P.: 1944, *Am. Phil. Soc.* **88**, 372.
Kamp, van de, P.: 1945, *Astron. J.* **51**, 161.
Kamp, van de, P.: 1947a, *Astron. J.* **52**, 185, 189.
Kamp, van de, P.: 1947b, *Astron. J.* **53**, 207.
Kamp, van de, P.: 1951, *Pop. Astron.* **51**, 65, 129, 176, 243.
Kamp, van de, P.: 1956, *Vistas Astron.* **2**, 1040.
Kamp, van de, P.: 1962, *J. Roy. Astron. Soc. Canada* **56**, 247.
Kamp, van de, P.: 1963, *Astron. J.* **68**, 515.
Kamp, van de, P.: 1964, *Elements of Astromechanics*, W. H. Freeman and Co., San Francisco, p. 129.
Kamp, van de, P.: 1965, *Publ. Astron. Soc. Pacific* **77**, 325.
Kamp, van de, P.: 1967, *Principles of Astrometry*, W. H. Freeman and Co., San Francisco.
Kamp, van de, P.: 1969a, *Astron. J.* **74**, 238.
Kamp, van de, P.: 1969b, *Astron. J.* **74**, 757.
Kamp, van de, P.: 1971, *IAU Symp.* **42**, 32.
Kamp, van de, P.: 1973, *Astron. J.* **78**, 1099.
Kamp, van de, P.: 1975, *Astron. J.* **80**. 658.
Kamp, van de, P.: 1976, *Royal Greenwich Observatory Bull.* **182**, 7.
Kamp, van de, P.: 1977a, *Vistas Astron.* **20**, 501.
Kamp, van de, P.: 1977b, *Astron. J.* **82**, 750.
Kamp, van de, P.: 1977c, *Vistas Astron.* **21**, 289.
Kamp, van de, P.: 1978a, *Astron. J.* **83**, 975.
Kamp, van de, P.: 1978b, *Sky Telesc.* **56**, 397.
Kamp, van de, P.: 1979, *Sky Telesc.* **57**, 247.
Kamp, van de, P.: 1980, *Astronomie* **94**, 207.
Kamp, van de, P. and Lippincott, S. L.: 1949, *Astron. J.* **55**, 16.
Kamp, van de, P. and Lippincott, S. L.: 1950, *Astron. J.* **55**, 89.
Kamp, van de, P. and Worth, M. D.: 1972, *Astron. J.* **77**, 762.
Kamp, van de, P. and Lippincott, S. L.: 1980, *Bull. Am. Astron. Soc.* **12**, 455.

Kamper, K. W.: 1971, *Publ. Leander McCormick Obs.* **16**, 285.
Kamper, K. W. and Wesselink, A. J.: 1978, *Astron. J.* **83**, 1653.
Kapteyn, J. C.: 1904, Address before the St. Louis Exposition Congress.
Kapteyn, J. C.: 1911, *First Report on the Progress of the Plan of Selected Areas*, p. 12.
Kapteyn, J. C.: 1922, *Bull. Astron. Inst. Netherlands*, No. 14, 71.
Land, G.: 1944, *Astron. J.* **51**, 25.
Lindblad, B.: 1926, *Arkiv Math. Astron. Fysik*, **19A**, Nos 21, 27; **19B**, No. 7.
Lindblad, B.: 1927, *Arkiv Math. Astron. Fysik* **20A**, No. 35.
Lippincott, S. L.: 1951, *Astron. J.* **55**, 236.
Lippincott, S. L.: 1953, *Astron. J.* **58**, 135.
Lippincott, S. L.: 1955a, *Sky Telesc.* **14**, 364.
Lippincott, S. L.: 1955b, *Astron. J.* **60**, 379.
Lippincott, S. L.: 1957, *Astron. J.* **62**, 55.
Lippincott, S. L.: 1965, *Vistas Astron.* **6**, 125.
Lippincott, S. L.: 1967, *Astron. J.* **72**, 1349.
Lippincott, S. L.: 1971, *Publ. Leander McCormick Obs.* **16**, 59.
Lippincott, S. L.: 1973, *Astron. J.* **78**, 426.
Lippincott, S. L.: 1974, *Astron. J.* **79**, 974.
Lippincott, S. L.: 1977, *Astron. J.* **82**, 925.
Lippincott, S. L.: 1978, *Space Sci. Rev.* **22**, 153.
Lippincott, S. L.: 1979, *Publ. Astron. Soc. Pacific*, **91**, 784.
Lippincott, S. L.: 1980, *Bull. Am. Astron. Soc.* **12**, 453.
Lippincott, S. L. and Hershey, J. L.: 1972, *Astron. J.* **77**, 679.
Lippincott, S. L. and Borgman, E. R.: 1978, *Publ. Astron. Soc. Pacific* **90**, 226.
Lippincott, S. L. and Lanning, J. J.: 1976, *Bull. Am. Astron. Soc.* **8**, 360.
Lippincott, S. L. and Turner, A.: 1980, *Publ. Astron. Soc. Pacific* **92**, 531.
Lippincott, S. L. and Worth, M. D.: 1978, *Publ. Astron. Soc. Pacific* **90**, 330.
Lundmark, K. and Luyten, W. J.: 1922, *Publ. Astron. Soc. Pacific* **34**, 126, 229.
Luyten, W. J.: 1968, *Monthly Notices Roy. Astron. Soc.* **139**, 22.
Martin, G. E. and Ianna, P. A.: 1975, *Astron. J.* **80**, 821.
McAlister, H. A.: 1977, *Sky Telesc.* **53**, 346.
Morel, P. J.: 1970, *Astron. Astrophys.* **6**, No. 3, 441.
Morris, S. C.: 1963, Thesis, Univ. Toronto.
Nörlund, N. E.: 1905, *Astron. Nachr.* **170**, 117.
Oort, J. H.: 1922, *Bull. Astron. Inst. Netherlands* **1**, 134.
Oort, J. H.: 1927, *Bull. Astron. Inst. Netherlands* **3**, 275.
Oort, J. H.: 1932, *Bull Astron. Inst. Netherlands* **6**, 87.
Oort, J. H.: 1965, *Galactic Structure*, Univ. of Chicago Press, p. 470.
Oort, J. H.: 1976, *Royal Greenwich Observatory Bull.* **182**, 11.
Payne-Gaposchkin, C.: 1941, *Publ. Am. Astron. Soc.* **10**, 127.
Pickering, E. C.: 1916, *Harvard Bulletin*, No. 613.
Probst, R. G.: 1977, *Astron. J.* **82**, 656.
Reuyl, D.: 1936, *Astron. J.* **45**, 133.
Ristenpart, F.: 1902, *Vierteljahrschrift* **37**, 342.
Russell, H. N.: 1898, *Astron. J.* **19**, 9.
Russell, H. N. and Atkinson, R. E.: 1931, *Nature* **137**, 661.
Schlesinger, F.: 1910, *Astrophys J.* **32**, 372.
Schlesinger, F.: 1911, *Astrophys. J.* **33**, 8.

Schlesinger, F.: 1917, *Astron. J.* **30**, 137.
Schlesinger, F.: 1924, *Probleme der Astronomie*, Springer Verlag, p. 422.
Schwarzschild, K.: 1907, *Göttinger Nachr.*, p. 614.
Schwarzschild, K.: 1908, *Göttinger Nachr.*, p. 191.
Seeliger, H.: 1888, *Abh. d. Münchener Akademie*, II Cl., Band XVII, 1. Abt.
Seeliger, H.: 1901, *Astron. Nachr.* **154**, 67.
Seeliger, H.: 1914, *Astron. Nachr.* **199**, 273.
Shapiro, S. L. and Teukolsky, S. A.: 1976, *Astrophys. J.* **203**, 697.
Shapley, H.: 1918, *Astrophys. J.* **48**, 89, 154; Mount Wilson Contr. Nos. 151, 152.
Shapley, H.: 1919, *Astrophys. J.* **49**, 249, 311; Mount Wilson Contr. Nos. 156, 157.
Smart, W. M.: 1960, *Spherical Astronomy*, p. 351.
Straka, W. C.: 1971, *Astrophys. J.* **165**, 109.
Strand, K. Aa.: 1937, *Leiden Annals* **18**, pt. 2
Strand, K. Aa.: 1942, *Astron. J.* **49**, 165.
Strand, K. Aa.: 1946, *Astron. J.* **52**, 1.
Strand, K. Aa: 1964, *Science* **144**, No. 3624, 1299.
Strand, K. Aa.: 1971, *Publ. U.S. Naval Observatory, Sec. Series*, Vol. XX Part 1.
Strand, K. Aa.: 1977, *Astron. J.* **82**, 745.
Strömberg, G. B.: 1925, *Astrophys. J.* **16**, 363; Mount Wilson Contr. No. 293.
Strong, J.: 1958, *Concepts of Classical Optics*, W. H. Freeman and Co.
Trumpler, R. J.: 1930, *Lick Observatory Bull.* **14**, 154.
Wanner, J. F.: 1967, *Sky Telesc.* **33**, No. 1, 16.
Wanner, J. F.: 1968, *Rev. Pop. Astron.* **62**, No. 550, 11.
Wright, K. O.: 1970, *Vistas Astron.* **12**, 150.
Wright, W. H.: 1910, *Astrophys. J.* **12**, 254.
Young, C. A.: 1904, *General Astronomy*, Ginn and Company, p. 414.
Zwiers, H. F.: 1895, *Astron. Nachr.* **139**, 369.

INDEX

ASTROPHYSICS AND SPACE SCIENCE LIBRARY

Edited by

J. E. Blamont, R. L. F. Boyd, L. Goldberg, C. de Jager, Z. Kopal, G. H. Ludwig, R. Lüst, B. M. McCormac, H. E. Newell, L. I. Sedov, Z. Švestka, and W. de Graaff

1. C. de Jager (ed.), *The Solar Spectrum, Proceedings of the Symposium held at the University of Utrecht, 26–31 August, 1963.* 1965, XIV + 417 pp.
2. J. Orthner and H. Maseland (eds.), *Introduction to Solar Terrestrial Relations, Proceedings of the Summer School in Space Physics held in Alpbach, Austria, July 15–August 10, 1963 and Organized by the European Preparatory Commission for Space Research.* 1965, IX + 506 pp.
3. C. C. Chang and S. S. Huang (eds.), *Proceedings of the Plasma Space Science Symposium, held at the Catholic University of America, Washington, D.C., June 11–14, 1963.* 1965, IX + 377 pp.
4. Zdeněk Kopal, *An Introduction to the Study of the Moon.* 1966, XII + 464 pp.
5. B. M. McCormac (ed.), *Radiation Trapped in the Earth's Magnetic Field. Proceedings of the Advanced Study Institute, held at the Chr. Michelsen Institute, Bergen, Norway, August 16–September 3, 1965.* 1966, XII + 901 pp.
6. A. B. Underhill, *The Early Type Stars.* 1966, XII + 282 pp.
7. Jean Kovalevsky, *Introduction to Celestial Mechanics.* 1967, VIII + 427 pp.
8. Zdeněk Kopal and Constantine L. Goudas (eds.), *Measure of the Moon. Proceedings of the 2nd International Conference on Selenodesy and Lunar Topography, held in the University of Manchester, England, May 30–June 4, 1966.* 1967, XVIII + 479 pp.
9. J. G. Emming (ed.), *Electromagnetic Radiation in Space. Proceedings of the 3rd ESRO Summer School in Space Physics, held in Alpbach, Austria, from 19 July to 13 August, 1965.* 1968, VIII + 307 pp.
10. R. L. Carovillano, John F. McClay, and Henry R. Radoski (eds.), *Physics of the Magnetosphere, Based upon the Proceedings of the Conference held at Boston College, June 19–28, 1967.* 1968, X + 686 pp.
11. Syun-Ichi Akasofu, *Polar and Magnetospheric Substorms.* 1968, XVIII + 280 pp.
12. Peter M. Millman (ed.), *Meteorite Research. Proceedings of a Symposium on Meteorite Research, held in Vienna, Austria, 7–13 August, 1968.* 1969, XV + 941 pp.
13. Margherita Hack (ed.), *Mass Loss from Stars. Proceedings of the 2nd Trieste Colloquium on Astrophysics, 12–17 September, 1968.* 1969, XII + 345 pp.
14. N. D'Angelo (ed.), *Low-Frequency Waves and Irregularities in the Ionosphere. Proceedings of the 2nd ESRIN-ESLAB Symposium, held in Frascati, Italy, 23–27 September, 1968.* 1969, VII + 218 pp.
15. G. A. Partel (ed.), *Space Engineering. Proceedings of the 2nd International Conference on Space Engineering, held at the Fondazione Giorgio Cini, Isola di San Giorgio, Venice, Italy, May 7–10, 1969.* 1970, XI + 728 pp.
16. S. Fred Singer (ed.), *Manned Laboratories in Space. Second International Orbital Laboratory Symposium.* 1969, XIII + 133 pp.
17. B. M. McCormac (ed.), *Particles and Fields in the Magnetosphere. Symposium Organized by the Summer Advanced Study Institute, held at the University of California, Santa Barbara, Calif., August 4–15, 1969.* 1970, XI + 450 pp.
18. Jean-Claude Pecker, *Experimental Astronomy.* 1970, X + 105 pp.
19. V. Manno and D. E. Page (eds.), *Intercorrelated Satellite Observations related to Solar Events. Proceedings of the 3rd ESLAB/ESRIN Symposium held in Noordwijk, The Netherlands, September 16–19, 1969.* 1970, XVI + 627 pp.
20. L. Mansinha, D. E. Smylie, and A. E. Beck, *Earthquake Displacement Fields and the Rotation of the Earth, A NATO Advanced Study Institute Conference Organized by the Department of Geophysics, University of Western Ontario, London, Canada, June 22–28, 1969.* 1970, XI + 308 pp.
21. Jean-Claude Pecker, *Space Observatories.* 1970, XI + 120 pp.
22. L. N. Mavridis (ed.), *Structure and Evolution of the Galaxy. Proceedings of the NATO Advanced Study Institute, held in Athens, September 8–19, 1969.* 1971, VII + 312 pp.

23. A. Muller (ed.), *The Magellanic Clouds. A European Southern Observatory Presentation: Principal Prospects, Current Observational and Theoretical Approaches, and Prospects for Future Research, Based on the Symposium on the Magellanic Clouds, held in Santiago de Chile, March 1969, on the Occasion of the Dedication of the European Southern Observatory.* 1971, XII + 189 pp.
24. B. M. McCormac (ed.), *The Radiating Atmosphere. Proceedings of a Symposium Organized by the Summer Advanced Study Institute, held at Queen's University, Kingston, Ontario, August 3–14, 1970.* 1971, XI + 455 pp.
25. G. Fiocco (ed.), *Mesospheric Models and Related Experiments. Proceedings of the 4th ESRIN-ESLAB Symposium, held at Frascati, Italy, July 6–10, 1970.* 1971, VIII + 298 pp.
26. I. Atanasijević, *Selected Exercises in Galactic Astronomy.* 1971, XII + 144 pp.
27. C. J. Macris (ed.), *Physics of the Solar Corona. Proceedings of the NATO Advanced Study Institute on Physics of the Solar Corona, held at Cavouri-Vouliagmeni, Athens, Greece, 6–17 September 1970.* 1971, XII + 345 pp.
28. F. Delobeau, *The Environment of the Earth.* 1971, IX + 113 pp.
29. E. R. Dyer (general ed.), *Solar-Terrestrial Physics/1970. Proceedings of the International Symposium on Solar-Terrestrial Physics, held in Leningrad, U.S.S.R., 12–19 May 1970.* 1972, VIII + 938 pp.
30. V. Manno and J. Ring (eds.), *Infrared Detection Techniques for Space Research. Proceedings of the 5th ESLAB-ESRIN Symposium, held in Noordwijk, The Netherlands, June 8–11, 1971.* 1972, XII + 344 pp.
31. M. Lecar (ed.), *Gravitational N-Body Problem. Proceedings of IAU Colloquium No. 10, held in Cambridge, England, August 12–15, 1970.* 1972, XI + 441 pp.
32. B. M. McCormac (ed.), *Earth's Magnetospheric Processes. Proceedings of a Symposium Organized by the Summer Advanced Study Institute and Ninth ESRO Summer School, held in Cortina, Italy, August 30–September 10, 1971.* 1972, VIII + 417 pp.
33. Antonin Rükl, *Maps of Lunar Hemispheres.* 1972, V + 24 pp.
34. V. Kourganoff, *Introduction to the Physics of Stellar Interiors.* 1973, XI + 115 pp.
35. B. M. McCormac (ed.), *Physics and Chemistry of Upper Atmospheres. Proceedings of a Symposium Organized by the Summer Advanced Study Institute, held at the University of Orléans, France, July 31–August 11, 1972.* 1973, VIII + 389 pp.
36. J. D. Fernie (ed.), *Variable Stars in Globular Clusters and in Related Systems. Proceedings of the IAU Colloquium No. 21, held at the University of Toronto, Toronto, Canada, August 29–31, 1972.* 1973, IX + 234 pp.
37. R. J. L. Grard (ed.), *Photon and Particle Interaction with Surfaces in Space. Proceedings of the 6th ESLAB Symposium, held at Noordwijk, The Netherlands, 26–29 September, 1972.* 1973, XV + 577 pp.
38. Werner Israel (ed.), *Relativity, Astrophysics and Cosmology. Proceedings of the Summer School, held 14–26 August, 1972, at the BANFF Centre, BANFF, Alberta, Canada.* 1973, IX + 323 pp.
39. B. D. Tapley and V. Szebehely (eds.), *Recent Advances in Dynamical Astronomy. Proceedings of the NATO Advanced Study Institute in Dynamical Astronomy, held in Cortina d'Ampezzo, Italy, August 9–12, 1972.* 1973, XIII + 468 pp.
40. A. G. W. Cameron (ed.), *Cosmochemistry. Proceedings of the Symposium on Cosmochemistry, held at the Smithsonian Astrophysical Observatory, Cambridge, Mass., August 14–16, 1972.* 1973, X + 173 pp.
41. M. Golay, *Introduction to Astronomical Photometry.* 1974, IX + 364 pp.
42. D. E. Page (ed.), *Correlated Interplanetary and Magnetospheric Observations. Proceedings of the 7th ESLAB Symposium, held at Saulgau, W. Germany, 22–25 May, 1973.* 1974, XIV + 662 pp.
43. Riccardo Giacconi and Herbert Gursky (eds.), *X-Ray Astronomy.* 1974, X + 450 pp.
44. B. M. McCormac (ed.), *Magnetospheric Physics. Proceedings of the Advanced Summer Institute, held in Sheffield, U.K., August 1973.* 1974, VII + 399 pp.
45. C. B. Cosmovici (ed.), *Supernovae and Supernova Remnants. Proceedings of the International Conference on Supernovae, held in Lecce, Italy, May 7–11, 1973.* 1974, XVII + 387 pp.
46. A. P. Mitra, *Ionospheric Effects of Solar Flares.* 1974, XI + 294 pp.
47. S.-I. Akasofu, *Physics of Magnetospheric Substorms.* 1977, XVIII + 599 pp.

48. H. Gursky and R. Ruffini (eds.), *Neutron Stars, Black Holes and Binary X-Ray Sources*. 1975, XII + 441 pp.
49. Z. Švestka and P. Simon (eds.), *Catalog of Solar Particle Events 1955–1969. Prepared under the Auspices of Working Group 2 of the Inter-Union Commission on Solar-Terrestrial Physics*. 1975, IX + 428 pp.
50. Zdeněk Kopal and Robert W. Carder, *Mapping of the Moon*. 1974, VIII + 237 pp.
51. B. M. McCormac (ed.), *Atmospheres of Earth and the Planets. Proceedings of the Summer Advanced Study Institute, held at the University of Liège, Belgium, July 29–August 8, 1974*. 1975, VII + 454 pp.
52. V. Formisano (ed.), *The Magnetospheres of the Earth and Jupiter. Proceedings of the Neil Brice Memorial Symposium, held in Frascati, May 28–June 1, 1974*. 1975, XI + 485 pp.
53. R. Grant Athay, *The Solar Chromosphere and Corona: Quiet Sun*. 1976, XI + 504 pp.
54. C. de Jager and H. Nieuwenhuijzen (eds.), *Image Processing Techniques in Astronomy. Proceedings of a Conference, held in Utrecht on March 25–27, 1975*. XI + 418 pp.
55. N. C. Wickramasinghe and D. J. Morgan (eds.), *Solid State Astrophysics. Proceedings of a Symposium, held at the University College, Cardiff, Wales, 9–12 July 1974*. 1976, XII + 314 pp.
56. John Meaburn, *Detection and Spectrometry of Faint Light*. 1976, IX + 270 pp.
57. K. Knott and B. Battrick (eds.), *The Scientific Satellite Programme during the International Magnetospheric Study. Proceedings of the 10th ESLAB Symposium, held at Vienna, Austria, 10–13 June 1975*. 1976, XV + 464 pp.
58. B. M. McCormac (ed.), *Magnetospheric Particles and Fields. Proceedings of the Summer Advanced Study School, held in Graz, Austria, August 4–15, 1975*. 1976, VII + 331 pp.
59. B. S. P. Shen and M. Merker (eds.), *Spallation Nuclear Reactions and Their Applications*. 1976, VIII + 235 pp.
60. Walter S. Fitch (ed.), *Multiple Periodic Variable Stars. Proceedings of the International Astronomical Union Colloquium No. 29, held at Budapest, Hungary, 1–5 September 1976*. 1976, XIV + 348 pp.
61. J. J. Burger, A. Pedersen, and B. Battrick (eds.), *Atmospheric Physics from Spacelab. Proceedings of the 11th ESLAB Symposium, Organized by the Space Science Department of the European Space Agency, held at Frascati, Italy, 11–14 May 1976*. 1976, XX + 409 pp.
62. J. Derral Mulholland (ed.), *Scientific Applications of Lunar Laser Ranging. Proceedings of a Symposium held in Austin, Tex., U.S.A., 8–10 June, 1976*. 1977, XVII + 302 pp.
63. Giovanni G. Fazio (ed.), *Infrared and Submillimeter Astronomy. Proceedings of a Symposium held in Philadelphia, Penn., U.S.A., 8–10 June, 1976*. 1977, X + 226 pp.
64. C. Jaschek and G. A. Wilkins (eds.), *Compilation, Critical Evaluation and Distribution of Stellar Data. Proceedings of the International Astronomical Union Colloquium No. 35, held at Strasbourg, France, 19–21 August, 1976*. 1977, XIV + 316 pp.
65. M. Friedjung (ed.), *Novae and Related Stars. Proceedings of an International Conference held by the Institut d'Astrophysique, Paris, France, 7–9 September, 1976*. 1977, XIV + 228 pp.
66. David N. Schramm (ed.), *Supernovae. Proceedings of a Special IAU-Session on Supernovae held in Grenoble, France, 1 September, 1976*. 1977, X + 192 pp.
67. Jean Audouze (ed.), *CNO Isotopes in Astrophysics. Proceedings of a Special IAU Session held in Grenoble, France, 30 August, 1976*. 1977, XIII + 195 pp.
68. Z. Kopal, *Dynamics of Close Binary Systems*, XIII + 510 pp.
69. A. Bruzek and C. J. Durrant (eds.), *Illustrated Glossary for Solar and Solar-Terrestrial Physics*. 1977, XVIII + 204 pp.
70. H. van Woerden (ed.), *Topics in Interstellar Matter*. 1977, VIII + 295 pp.
71. M. A. Shea, D. F. Smart, and T. S. Wu (eds.), *Study of Travelling Interplanetary Phenomena*. 1977, XII + 439 pp.
72. V. Szebehely (ed.), *Dynamics of Planets and Satellites and Theories of Their Motion. Proceedings of IAU Colloquium No. 41, held in Cambridge, England, 17–19 August 1976*. 1978, XII + 375 pp.
73. James R. Wertz (ed.), *Spacecraft Attitude Determination and Control*. 1978, XVI + 858 pp.

74. Peter J. Palmadesso and K. Papadopoulos (eds.), *Wave Instabilities in Space Plasmas. Proceedings of a Symposium Organized Within the XIX URSI General Assembly held in Helsinki, Finland, July 31–August 8, 1978.* 1979, VII + 309 pp.
75. Bengt E. Westerlund (ed.), *Stars and Star Systems. Proceedings of the Fourth European Regional Meeting in Astronomy held in Uppsala, Sweden, 7–12 August, 1978.* 1979, XVIII + 264 pp.
76. Cornelis van Schooneveld (ed.), *Image Formation from Coherence Functions in Astronomy. Proceedings of IAU Colloquium No. 49 on the Formation of Images from Spatial Coherence Functions in Astronomy, held at Groningen, The Netherlands, 10–12 August 1978.* 1979, XII + 338 pp.
77. Zdeněk Kopal, *Language of the Stars. A Discourse on the Theory of the Light Changes of Eclipsing Variables.* 1979, VIII + 280 pp.
78. S.-I. Akasofu (ed.), *Dynamics of the Magnetosphere. Proceedings of the A.G.U. Chapman Conference 'Magnetospheric Substorms and Related Plasma Processes' held at Los Alamos Scientific Laboratory, N.M., U.S.A., October 9-13, 1978.* 1980, XII + 658 pp.
79. Paul S. Wesson, *Gravity, Particles, and Astrophysics. A Review of Modern Theories of Gravity and G-variability, and their Relation to Elementary Particle Physics and Astrophysics.* 1980, VIII + 188 pp.
80. Peter A. Shaver (ed.), *Radio Recombination Lines. Proceedings of a Workshop held in Ottawa, Ontario, Canada, August 24-25, 1979.* 1980, X + 284 pp.
81. Pier Luigi Bernacca and Remo Ruffini (eds.), *Astrophysics from Spacelab*, 1980, XI + 664 pp.
82. Hannes Alfvén, *Cosmic Plasma*, 1981, X + 160 pp.
83. Michael D. Papagiannis (ed.), *Strategies for the Search for Life in the Universe*, 1980, XVI + 254 pp.
84. H. Kikuchi (ed.), *Relation between Laboratory and Space Plasmas*, 1981, XII + 386 pp.
85. Peter van der Kamp, *Stellar Paths*, 1981, XXII + 155 pp.
86. E. M. Gaposchkin and B. Kołaczek (eds.), *Reference Coordinate Systems for Earth Dynamics*, 1981, XIV + 396 pp.
87. R. Giacconi (ed.), *X-Ray Astronomy with the Einstein Satellite. Proceedings of the High Energy Astrophysics Division of the American Astronomical Society Meeting on X-Ray Astronomy held at the Harvard-Smithsonian Center for Astrophysics, Cambridge, Mass., U.S.A., January 28–30, 1980.* 1981, VII + 330 pp.
88. Icko Iben Jr. and Alvio Renzini (eds.), *Physical Processes in Red Giants. Proceedings of the Second Workshop, held at the Ettore Majorana Centre for Scientific Culture, Advanced School of Agronomy, in Erice, Siciliy, Italy, September 3–13, 1980.* 1981, XV + 488 pp.
89. C. Chiosi and R. Stalio (eds.), *Effect of Mass Loss on Stellar Evolution. IAU Colloquium No. 59 held in Miramare, Trieste, Italy, September 15–19, 1980.* 1981, XXII + 532 pp.
90. C. Goudis, *The Orion Complex: A Case Study of Interstellar Matter*, 1982 (forthcoming).
91. F. D. Kahn (ed.), *Investigating the Universe. Papers Presented to Zdeněk Kopal on the Occasion of his retirement, September 1981.* 1981, X + 458 pp.

GPSR Compliance

The European Union's (EU) General Product Safety Regulation (GPSR) is a set of rules that requires consumer products to be safe and our obligations to ensure this.

If you have any concerns about our products, you can contact us on ProductSafety@springernature.com

In case Publisher is established outside the EU, the EU authorized representative is:

Springer Nature Customer Service Center GmbH
Europaplatz 3
69115 Heidelberg, Germany

Batch number: 10607224

Printed by Printforce, the Netherlands